Collector's Guide to the
AXINITE GROUP

Schiffer Earth Science Monographs Volume 4

Robert J. Lauf

Other Schiffer Books by Robert J. Lauf
Collector's Guide to the Epidote Group. Robert J. Lauf.
ISBN: 9780764330483. $19.99.
Collector's Guide to the Mica Group. Robert J. Lauf.
ISBN: 9780764330476. $19.99.
Introduction to Radioactive Minerals. Robert J. Lauf.
ISBN: 9780764329128. $29.95.

Other Schiffer Books on Related Subjects
Collecting Fluorescent Minerals. Stuart Schneider.
ISBN: 0764320912. $29.95.
Collector's Guide to the Fluorite Group. Arvid Eric Pasto.
ISBN: 978-0-7643-3193-0. $19.99.
Gems & Minerals. Dr. Andreas Landmann.
ISBN: 9780764330667. $29.99.
The World of Fluorescent Minerals. Stuart Schneider.
ISBN: 0764325442. $29.95.

Copyright © 2009 by Robert J. Lauf
Library of Congress Control Number: 2008940631

All rights reserved. No part of this work may be reproduced or used in any form or by any means—graphic, electronic, or mechanical, including photocopying or information storage and retrieval systems—without written permission from the publisher.

The scanning, uploading and distribution of this book or any part thereof via the Internet or via any other means without the permission of the publisher is illegal and punishable by law. Please purchase only authorized editions and do not participate in or encourage the electronic piracy of copyrighted materials.

"Schiffer," "Schiffer Publishing Ltd. & Design," and the "Design of pen and ink well" are registered trademarks of Schiffer Publishing Ltd.

Designed by Mark David Bowyer
Type set in Arno Pro / Humanist521 BT

ISBN: 978-0-7643-3216-6
Printed in China

Schiffer Books are available at special discounts for bulk purchases for sales promotions or premiums. Special editions, including personalized covers, corporate imprints, and excerpts can be created in large quantities for special needs. For more information contact the publisher:

Published by Schiffer Publishing Ltd.
4880 Lower Valley Road
Atglen, PA 19310
Phone: (610) 593-1777; Fax: (610) 593-2002
E-mail: Info@schifferbooks.com

For the largest selection of fine reference books on this and related subjects, please visit our web site at
www.schifferbooks.com
We are always looking for people to write books on new and related subjects. If you have an idea for a book please contact us at the above address.

This book may be purchased from the publisher.
Include $5.00 for shipping.
Please try your bookstore first.
You may write for a free catalog.

In Europe, Schiffer books are distributed by
Bushwood Books
6 Marksbury Ave.
Kew Gardens
Surrey TW9 4JF England
Phone: 44 (0) 20 8392-8585; Fax: 44 (0) 20 8392-9876
E-mail: info@bushwoodbooks.co.uk
Website: www.bushwoodbooks.co.uk
Free postage in the U.K., Europe; air mail at cost.

Contents

Preface 5

Acknowledgments 7

Introduction 8

Taxonomy of the Axinite Group 13
    General formula 13
    Crystal structure and morphology 21

Formation and Geochemistry 28
    Axinites in igneous rocks 28
    Axinites in metamorphic rocks 30

The Minerals 41
    Axinite-(Fe) 41
    Axinite-(Mg) 66
    Axinite-(Mn) 71
    Tinzenite 88

References 91

Preface

This volume continues a series of monographs on important groups of so-called rock forming silicates, the purpose of which is to help mineral collectors gain a better appreciation of these complex minerals. Because of the importance of rock forming minerals in geological processes, they are the subject of extensive published research, much of which has been brought together in the five-volume compendium *Rock-Forming Minerals* (Deer, Howie, and Zussman 1962) and the greatly expanded Second Edition thereof. Among rock-forming minerals, the axinite group, although small, is well known to collectors through the spectacularly sharp crystals that have been collected in quantity at Dal'negorsk and Puiva in Russia and at Tomas and in the Kharan district in Pakistan. Axinites are occasionally cut as gems, and indeed the mineral axinite-(Mg) was first described (as magnesio-axinite) from a faceted stone in the collection of the British Museum.

The author considers it especially timely to take a detailed look at the axinite group because the nomenclature of the group has recently been addressed by a committee of the International Mineralogical Association (IMA) and the names of three of the four species have been changed (Burke 2008).

The present monograph, which grew out of a *Rocks & Minerals* paper by the author (Lauf 2007), is organized as follows: After a brief introduction, the general treatment begins with an explanation of the chemistry and taxonomy of the group. A section on their formation and geochemistry explains the kinds of environments where axinites are formed. Then, a detailed entry for each mineral provides information on notable localities and full-color photos wherever possible so that collectors can see what good specimens look like and which minerals one might expect to find in association with axinites.

Acknowledgments

The following colleagues kindly provided technical information, literature, and helpful discussions: Deborah Cole, *Oak Ridge National Laboratory;* Jan Filip, *Palacky University in Olomouc*; Arvid Pasto; Radek Škoda, *Masaryk University*. Important specimens and background information were supplied by Laurie Adams, *The Adams Collection*; Dudley Blauwet, *Mountain Minerals*; Dave Bunk; Sharon Cisneros, *Mineralogical Research Co.*; Richard Dale, *Dale Minerals*; Kevin Downey, *Well-Arranged Molecules*; Jordi Fabre, *www.fabreminerals.com*; Shields Flynn, *Trafford-Flynn Minerals*; Leonard Himes, *Minerals America*; Danny Jones, *Firebird Minerals*; Patrick Kelley, *PAK Designs*; Bill Logan, *Spectrum Minerals*; and Tony Nikischer, *Excalibur Mineral Co.*

Introduction

The axinite group comprises four triclinic borosilicate minerals known for their characteristically sharp, wedge-shaped or "axehead" crystals. They usually form in low- to medium-grade metamorphic environments, and in some regionally metamorphosed rocks axinites are the only borosilicate mineral. Axinites are found in many localities worldwide; in recent years, particularly fine specimens have been widely available from several localities in Russia and Pakistan.

As a gem material, axinite is occasionally faceted, although large stones are rare, and clean gems are generally less than about 5 carats. They tend to be strongly trichroic, with browns and purples predominating, and have the potential to make very beautiful gemstones. Axinite is hard enough to be safely worn in jewelry, but it tends to be somewhat brittle (Arem 1977). The strong pleochroism distinguishes axinite from smoky quartz, which is only slightly dichroic (Bauer 1904).

Figure 1. A single axinite crystal from Puiva, Russia, showing the sharp, bladed habit for which the mineral was named. *RJL1445*

Until the discovery of large crystals in Russia, the use of axinite as a gemstone was very limited. Some difficulties with the older material included the fact that many crystals were not sufficiently transparent, the colors often contained a lot of gray, and crystals tended to be small (and especially thin), which severely limited the size of stones that could be cut from them (Bauer 1904). Kunz (1892) commented, "No crystals of axinite have been found in this country of sufficient size to furnish gems. … Specimens from Dauphin, France, and Scopi, Switzerland, are occasionally cut into beautiful stone-brown gems, but for gem collections only."

Figure 2. Side view of the crystal in the previous figure, clearly showing the sharp edge and acute termination.

Axinite-(Mg) has the interesting distinction that the material was first "discovered" as a faceted gem. The type specimen is a round, pale blue mixed-cut gemstone weighing 0.78 carats. The stone is noticeably pleochroic, pale blue to pale violet to pale gray; it appears pale blue in daylight but pale violet under tungsten light. Ultraviolet fluorescence was noted as well: distinct orange-red under LW UV and duller red under SW UV (Jobbins, Tresham, and Young 1975).

In an attempt to simplify or rationalize mineral nomenclature, the International Mineralogical Association recently decided to change the names of many valid mineral species, generally tilting the nomenclature toward "Levinson notation," an approach originally used for rare-earth minerals, in which numerous rare earth elements can substitute for one another. Regrettably, all of the existing literature uses the "old" names, so the collector who wishes to consult traditional references must look for entries under the traditional names. Mineral specimens will no doubt continue to be labeled and sold under the old names for years to come, as well.

The impact of the IMA decision on the axinite group is as follows:

Ferro-axinite is now called **axinite-(Fe)**
Magnesio-axinite is now called **axinite-(Mg)**
Manganaxinite is now called **axinite-(Mn)**
Tinzenite is still called **tinzenite**.

Figure 3. Some faceted axinites. Left: two stones cut in different shapes from axinite-(Fe) from Pakistan; center top: a 5-carat round cut from axinite-(Fe) from the Puiva deposit, Russia; center bottom: a small pale lavender square-cut axinite-(Mg) from the Merelani region, Tanzania; right: two stones cut from axinite-(Fe) from King's Mountain, North Carolina.

Taxonomy of the Axinite Group

General Formula

The general formula and nomenclature for the axinite group (Sanero and Gottardi 1968) may be summarized as follows: $A_2MAl_2[BSi_4O_{15}](OH)$ where:

A is Ca or Mn^{2+};
M is Fe^{2+}, Mg, or Mn^{2+};

and the crystal structure is triclinic, with $P\bar{1}$ symmetry.

Most axinite analyses are very close to the nominal 2 Ca ions per formula unit. The structure of tinzenite is calcium-deficient, with some of the Ca being replaced by divalent Mn or Fe. Different authors have expressed the formulas in various equivalent or at least consistent ways; the idealized formulas for the four minerals, Table 1, emphasize their relationships and differences. Note that the unit cell contains two formula units.

As one might expect, the minerals form solid-solution series from one end member to another. Average compositions at several localities, plotted on the compositional triangle bounded by the respective Mg- Fe- and Mn-dominant end members, show that at some locales the mineral is close to an end member composition, e.g., the Tanzanian axinite-(Mg). In other deposits, the average composition lies closer to a boundary, seen, for example, in the axinite-(Fe) from Obira, Japan, where published analyses show a significant amount of Mn and at least one analysis actually falls within the axinite-(Mn) field. A particularly extreme case is seen at Luning, Nevada, where the average composition virtually straddles the boundary between axinite-(Fe) and axinite-(Mg). It is clear from the compositional triangle that when three different ions can occupy a particular site, the "dominant ion" might occupy less than 50 percent of the sites while still exceeding either of the other two ions. In other words, the "33 percent rule" applies rather than the "50 percent rule". For instance, the axinite-(Mg) from Luning, Nevada, plotted as data point **2** in the drawing, corresponds to $Mg_{0.42}Fe_{0.33}Mn_{0.26}$.

It has been observed that near-end-member axinite-(Fe), i.e., material with very little Mg or Mn, is fairly uncommon. This is attributed to the preferred incorporation of the relatively large Mn^{2+} ion at the **M** site, allowing axinite to act as a sink for Mn. Thus very "pure" axinite-(Fe) will only form when the surrounding rock or fluids are very low in Mn (Filip et al. 2006).

Table 1. Minerals of the axinite group

Name	Formula
Axinite-(Fe)	$Ca_2Fe^{2+}Al_2[BSi_4O_{15}](OH)$
Axinite-(Mg)	$Ca_2MgAl_2[BSi_4O_{15}](OH)$
Axinite-(Mn)	$Ca_2Mn^{2+}Al_2[BSi_4O_{15}](OH)$
Tinzenite	$CaMn^{2+}_2Al_2[BSi_4O_{15}](OH)$

Opposite Page:
Figure 4. Drawing of Mn-Mg-Fe compositional triangle showing some examples of the axinite compositions at various localities. Samples from some localities, such as the axinite-(Mg) from Arusha, Tanzania, axinite-(Mn) from Graham, Arizona, and axinite-(Fe) from Malešov, Czech Republic, are close to theoretical end-member composition. At other locales, compositions lie near a boundary between two species, such as the axinite-(Mg) from Luning, Nevada. Still other locales show significant compositional ranges that encompass two species, such as those found at Bourg d'Oisans, France, and Strigau, Poland. Data from Andreozzi et al. (2000) and other sources.

1 Arusha, Tanzania
2 Luning, Nevada
3 Bourg d'Oisans, France
4 Puiva, Russia
5 Sri Lanka
6 Malešov, Czech Republic
7 Hajikami, Japan
8 Striegau, Poland
9 Striegau, Poland
10 Bourg d'Oisans, France
11 Obira, Japan
12 Dal'negorsk, Russia
13 Dal'negorsk, Russia
14 Graham, Arizona

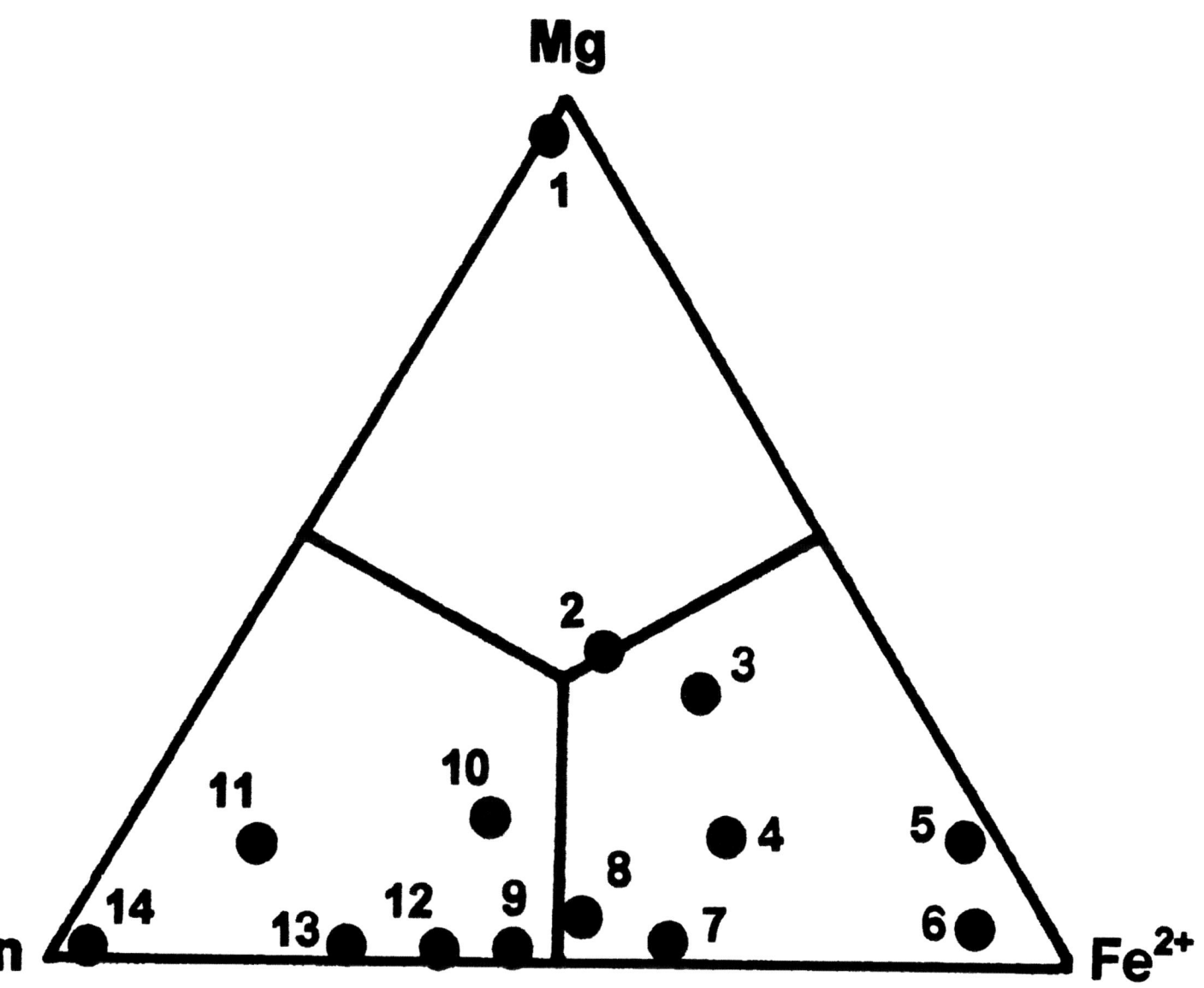

When minerals form a solid solution series, it is not uncommon for individual crystals to exhibit some compositional zoning, and axinites are no exception. Careful studies of zoned axinite crystals from many different localities indicate that the chemical zoning shows a fairly consistent trend from a more Fe-rich core to a more Mn-rich rim. Zoning is of two general types: The more common type shows a large Fe-rich core with a thin Mn-rich rim, whereas the second, less common type shows a very small Fe-rich core surrounded by a thick Mn-rich rim. The first type of zoning is usually seen in axinite-(Fe) and the second type in axinite-(Mn). In addition, (Fe, Mg)-rich samples show some depletion of Mg at the rim. One can easily see that in the situation where the average composition lies close to the 50:50 boundary it can well be the case that a single crystal might have an axinite-(Fe) core and an axinite-(Mn) rim (Andreozzi et al. 2000). Collectors will often come across fine specimens that are simply labeled "axinite." Properly identifying them as one of the accepted species can be a frustrating affair, because they can certainly not be reliably distinguished by eye or by any simple tests. Without recourse to microanalysis, the collector must make do with what information can be had from the literature, assuming the specimen is accurately labeled as to locality. Fortunately, axinites from many localities have been analyzed and several good summaries are available to help guide the identification process (Dunn, Leavens, and Barnes 1980; Deer, Howie, and Zussman 1986; Andreozzi et al. 2000).

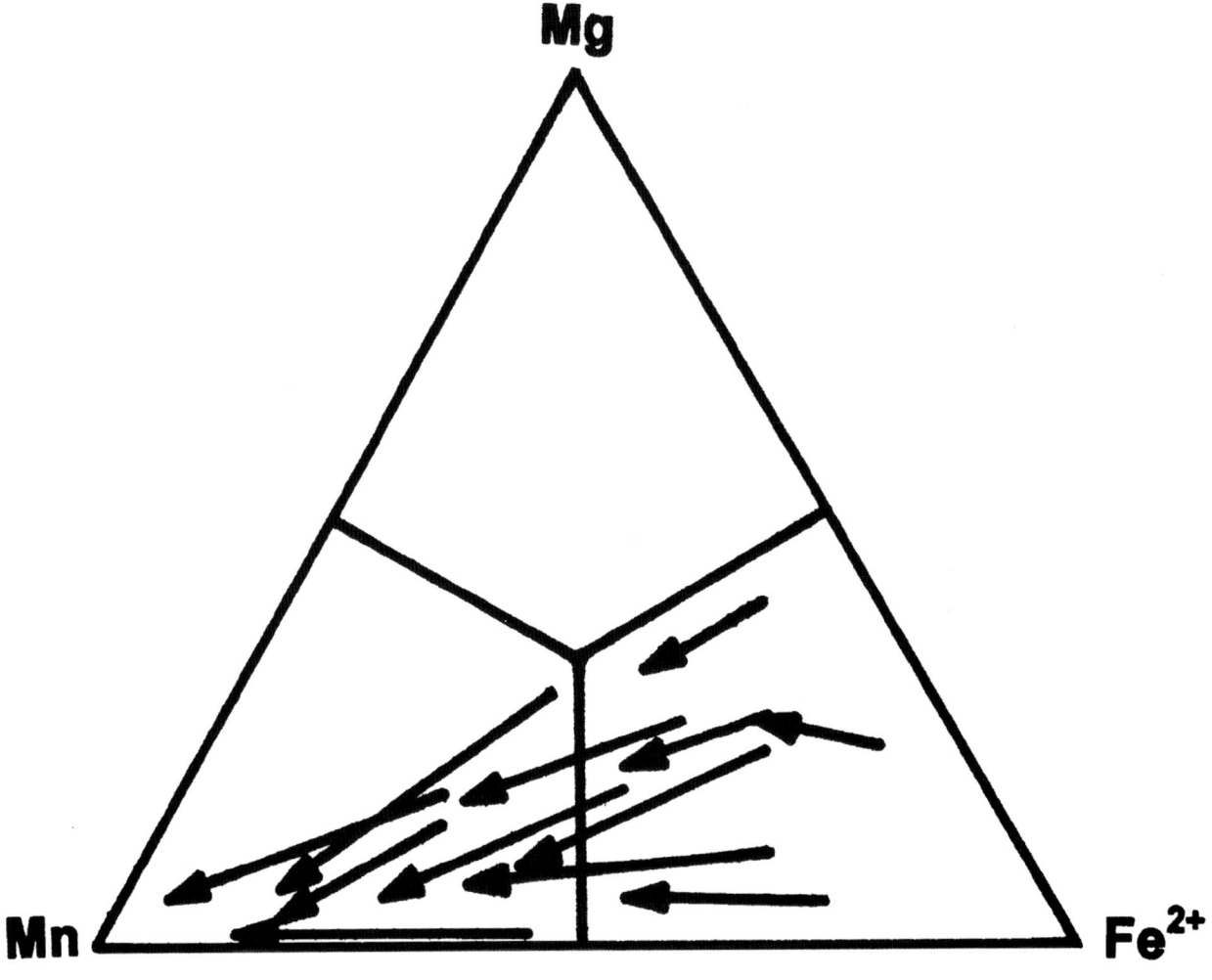

Figure 5. Drawing of Mn-Mg-Fe compositional triangle showing some examples of zoning in axinite single crystals. Each arrow covers the range of composition from core to rim for a single crystal, and it can be seen that in some cases the core is axinite-(Fe) and the rim is axinite-(Mn). Adapted from Andreozzi et al. (2000).

Published analyses are summarized in Table 2, organized by locality. There are occasional inconsistencies in the literature; for instance, material from the Huachuca Mountains in Arizona has been listed as tinzenite (Anthony et al. 1982) and as axinite-(Mn) (Dunn et al. 1980, based on a single analysis). Strictly speaking, each author's conclusions apply to the particular specimen(s) analyzed and as noted above two members of the group sometimes coexist in one crystal. The author has obtained several axinites that were only identified as axinite, have not been analyzed, and for which it was not possible to find reliable literature to identify the exact species; in these cases the specimen will simply be identified as "axinite". Collectors will also occasionally encounter specimens from the Pachapaqui district, Ancash department, Peru, labeled as either axinite-(Mn) or tinzenite, but look very similar and have similar associates. Axinite-(Mn) is listed as occurring in Pachapaqui by Crowley, Currier, and Szenics (1997) *but with a question mark*. As of this writing the author has not seen any published studies of these materials, and a systematic analytical study would be well worthwhile. The single analysis of Peruvian material presented by Lumpkin and Ribbe (1979), and interpreted as axinite-(Mn), is based on a sample that came from Caylloma, Arequipa department, Peru, hundreds of kilometers away. In this book, two specimens will be shown, with identifications [axinite-(Mn) and tinzenite, respectively] as provided by the original suppliers, with the caveat that the author regards the identifications as tentative.

To make matters even more interesting, recent structural work has raised the possibility (or threat, depending on one's viewpoint) of further nomenclatural changes, suggesting new terms such as "ferri-manganaxinite" and "ferri-ferroaxinite" to describe samples in which some Al is replaced by ferric iron. Earlier systematic analytical work on sixty axinites from twenty-four locales throughout the world (Andreozzi et al. 2000) had shown that B (rather than Al) is inversely related to Si and that Fe^{3+} mostly substitutes for Al but also for divalent cations, and these heterovalent substitutions are compensated by a deficiency in OH^-. Thus, to fully understand axinite crystal chemistry it is necessary to define the partitioning of Fe among tetrahedral and octahedral sites. To that end, nine samples were studied by X-ray diffraction, structural refinement, and ^{57}Fe Mössbauer spectroscopy in order to clarify the Fe distribution mechanisms and accompanying structural deformations. These studies showed that some samples do indeed have a significant amount of Fe^{3+} and ruled out the likelihood that Fe^{3+} can substitute for Si on tetrahedral sites. In B-deficient samples, it is Si (rather than Fe^{3+}) that substitutes for B, based on the fact that in all the analyzed samples the sum of B + Si was sufficient to fill all the tetrahedral sites. One of the two octahedral sites normally occupied by aluminum, **Z1**, is larger than the other site **Z2**. In one sample Fe^{3+} reached 13% occupancy of the **Z1** site; as it is conceivable that material might be found in which Fe^{3+} exceeds Al on **Z1**, terms such as ferri-manganaxinite and ferri-ferroaxinite were suggested (Andreozzi et al. 2004). At this writing it is not clear how that idea might play out in light of recent IMA decisions to implement Levinson-type suffixes and eliminate prefixes wherever possible (Burke 2008).

Table 2. Axinite analyses by locality

Locality	Mineral	Ref[a]
Australia		
Colebrook Hill, TAS	Axinite-(Fe)	DLB
Roseberry District, TAS	Axinite-(Fe)	And, LR
London Bridge, Queanbeyan, NSW	Axinite-(Mg)	DHZ
Austria		
Knappenwand, Tirol	Axinite-(Fe)	LR
Brazil		
Vitoria da Conquista	Axinite-(Fe)	DLB
Canada		
near Hope, BC	Axinite-(Fe)	DLB
Marmora Township, ONT	Axinite-(Mn)	DLB
Moneta mine, Timmins, ONT	Axinite-(Fe)	DLB
Czech Republic		
Davle, nr Praha	Axinite-(Fe)	DLB
Zbraslav, nr Praha	Axinite-(Fe)	DLB
England		
Botallack mine, St. Just, Cornwall	Axinite-(Fe)	DLB
Roscommon Cliff, St. Just, Cornwall	Axinite-(Fe)	LR
Carrick Dhu, St. Ives, Cornwall	Axinite-(Fe)	DHZ
Liskeard, Cornwall	Axinite-(Fe)	DLB
Tremore, Bodmin, Cornwall	Axinite-(Fe)	LR
Meldon, Okehampton, Devonshire	Axinite-(Fe)	DHZ
Finland		
Jolioinen	Axinite-(Fe)	DHZ
France		
Dauphin, Isere	Axinite-(Fe)	LR
St. Christophe-en-Oisans, Isere	Axinite-(Fe)	DLB
Bourg d'Oisans	Axinite-(Fe)	DHZ
Bourg d'Oisans	Axinite-(Mn)	And
Le Vernis, Bourg d'Oisans, Isere	Axinite-(Fe)	LR
Cornille, NW of Le Bourg d'Oisans	Axinite-(Fe)	DLB
Maison Guignard la Balme, Isere	Axinite-(Fe)	DLB
Pyrenees	Axinite-(Mn)	DLB
Germany		
Thum, Sachsen	Axinite-(Fe)	DLB
Ehrenfriedersdorf, Sachsen	Axinite-(Fe)	DLB
Pferdekopf, nr Wormke, Harz Mtns.	Axinite-(Fe)	LR
Freseburg, Harz Mtns.	Axinite-(Fe)	DLB
Saxony	Axinite-(Fe)	LR
Italy		
Cassagna mine, nr Chiavari, Genoa	Tinzenite	LR
Gambetesa, Campabasso	Tinzenite	DHZ

Locality	Mineral	Ref[a]
Gambatesa mine	Axinite-(Mn)	DHZ
Monzoni, Trentino-Alto Adige	Axinite-(Fe)	DLB
Tremola	Axinite-(Fe)	LR
Monte Pu, Liguria	Axinite-(Mn)	DHZ
S. Paolo Cervo, Biella	Axinite-(Fe)	And
Japan		
Fuji, nr Kamakura, Honshu	Axinite-(Fe)	DLB
Hajikami, Hyugo, Miazaki, Honshu	Axinite-(Fe)	And
Yamaura, Hyugo, Miazaki, Honshu	Axinite-(Fe)	DLB
Toroku mine, Iwato, Miazaki, Honshu	Axinite-(Fe)	LR
Iwatamura, Nishius, Oki gun, Hikoga	Axinite-(Fe)	DLB
Kuwanorizawa, Hikana, Tokyo, Honshu	Axinite-(Fe)	DLB
Obira, Bungo, Oita, Hyushu	Axinite-(Fe)[b]	DLB, LR
Takanosu mine, Kami Kasuo, Honshu	Axinite-(Mn)	DLB
Hata, Moji, Fukuoka Pref.	Axinite-(Fe)	DHZ
Anawai mine, Kochi-ken, Shikoku	Axinite-(Mn)	DLB
Mexico		
Trinidad, BC	Axinite-(Fe)	DLB
New Zealand		
Akatore Creek, South Is.	Tinzenite	PK
Dansey Pass, South Is.	Axinite-(Mn)	PK
Hawkdun Range, South Is.	Axinite-(Fe)	PK
Lower Pareora Gorge, South Is.	Axinite-(Mn)	PK
Malvern Hills, South Is.	Axinite-(Mn)	PK
Station Stream, South Is.	Axinite-(Mn)	PK
Norway		
Kongsberg, Buskerud	Axinite-(Fe)	LR
Peru		
Caylloma, Arequipa	Axinite-(Mn)	LR
Poland		
Striegau, Silesia	Axinite-(Mn)[d]	And
Russia		
Bakal, Southern Ural Mts.	Axinite-(Mg)	ES
Berkutshaja Sara Kreis, Zlatoust, Ural Mts.	Axinite-(Fe)	DLB
Dal'negorsk	Axinite-(Mn)	And
Puiva, Subpolar Urals	Axinite-(Fe)	And
Miask, Polar Urals	Axinite-(Fe)	And
Sri Lanka		
unknown	Axinite-(Fe)	And
Sweden		
Dannemora	Axinite-(Mn)	LR
Switzerland		
Alpe Parsettens, Val d'Err, nr Tinzen, Grisons	Tinzenite	DLB
Val d'Err, nr Tinzen	Tinzenite	LR
Tinzens, Grisons	Tinzenite	DHZ
Berg Sroyi, Grisons	Axinite-(Fe)	LR

Locality	Mineral	Ref[a]
Medels, Grisons	Axinite-(Fe)	DLB
Piz Vallatcha, nr Scopi, Grisons	Axinite-(Fe)	DLB
Santa Maria, nr Medels, Grisons	Axinite-(Fe)	DLB
Scopi, Mittelrheintal, Grisons	Axinite-(Fe)	LR
Sankt Jakob, Uri	Axinite-(Fe)	DLB
Tanzania		
Arusha district	Axinite-(Mg)	JTY
USA		
Graham, AZ	Axinite-(Mn)	And
Huachuca Mtns., AZ	Axinite-(Mn)	LR
Consummes Copper mine, Amador Co., CA	Axinite-(Mn)	DLB
Coarse Gold, Madera Co., CA	Axinite-(Fe)	DLB
Stinson Beach, Marin Co., CA	Axinite-(Fe)	DLB
Feather River, Plumas Co., CA	Axinite-(Fe)	LR
Greystone claim, Plumas Co., CA	Axinite-(Mn)	DLB
near Taylorville, Plumas Co., CA	Axinite-(Mn)	DLB
City quarry, Riverside Co., CA	Axinite-(Fe)	DLB
North Hill, Riverside Co., CA	Axinite-(Fe)	DLB
Bonsall, San Diego Co., CA	Axinite-(Fe)	LR
Klamath River, nr Yreka, Siskiyou Co., CA	Axinite-(Fe)	DLB
Woodlake, Tulare Co., CA	Axinite-(Mn)	LR
Thomaston Dam, Thomaston, CT	Axinite-(Fe)	LR
Biwabik iron formation, Mesabi Range, MN	Axinite-(Mn)	DHZ
Elkhorn, Jefferson Co., MT	Axinite-(Fe)	DLB
near Luning, NV	Axinite-(Mg)[c]	DLB
Mineral Co., NV	Axinite-(Fe)	LR
Franklin, Sussex Co., NJ	Axinite-(Mn)	DLB, LR
Foote mine, King's Mtn., NC	Axinite-(Fe)	DLB
McKinney mine, Mitchell Co., NC	Axinite-(Mn)	LR
Kibblehouse quarry, Perkiomenville, PA	Axinite-(Fe)	LR
Cornog, Chester Co., PA	Axinite-(Mn)	LR
near Campbell's Hump, PA	Axinite-(Mn)	LR
Avondale, Delaware Co., PA	Axinite-(Mn)	LR
near Easton, Northampton Co., PA	Axinite-(Mn)	DLB
Luck quarry, E of Charlottesville, VA	Axinite-(Fe)	LR

[a]And = Andreozzi et al. (2000); DLB = Dunn, Leavens, and Barnes (1980); DHZ = Deer, Howie, and Zussman (1986); ES = Erokhin and Shagalov (2007); JTY = Jobbins, Tresham, and Young (1975); LR = Lumpkin and Ribbe (1979); PK = Pringle and Kawachi (1980). Note: when more than one reference exists for a locality, preference has been given to those that actually report analytical data.

[b]One of nine analyses indicated axinite-(Mn)

[c]Five of seventeen analyses indicated axinite-(Fe)

[d]Analyses indicated some samples were axinite-(Fe)

Crystal Structure and Morphology

The triclinic crystal structure of axinite is generally described using the unit cell and orientation proposed by Peacock (1937), who demonstrated good agreement between the axial angles determined morphologically and the parameters derived from X-ray diffraction on the same crystal. Structure refinements showed that boron is tetrahedrally coordinated and the structure contains unique $B_2Si_8O_{30}$ groups in which two BO_4 tetrahedra share three corner oxygens each with neighboring SiO_4 tetrahedra, thereby linking together four Si_2O_7 groups. In the center of the structure there is a six-membered ring made up of two BO_4 and four SiO_4 tetrahedra. The $B_2Si_8O_{30}$ groups are nearly planar and arranged in tetrahedral layers alternating with sheets of edge-sharing octahedra. The groups lie approximately parallel to $(\bar{1}21)$, which is one of the characteristic forms (or faces) seen in axinite crystals (Ito et al. 1969).

Based on the structure refinement of a crystal from Woodlake, California, in which Fe:Mn ~ 1:1, there was good evidence that all of the Mn occupies the "iron" octahedral sites (**M** in the general formula given above) and this was believed to be the case for all axinite-(Mn) compositions up two Mn atoms per unit cell (Ito et al. 1969). When more manganese is present, the Mn will begin displacing Ca and at a high enough level the mineral becomes tinzenite. It is, however, unlikely that Mn could replace all of the Ca because this would cause the "Ca" coordination polyhedra to shrink so much that it would extensively distort the Si_2O_7 groups (Deer, Howie, and Zussman 1986).

Tinzenite was shown to be isostructural with axinite by Milton, Hildebrand, and Sherwood (1953), who also showed for the first time that tinzenite from the type locale contained boron, which was not noted in the analyses accompanying Jakob's original description. Later structure refinements on tinzenite from Cassagna, Italy (Basso, Della Guista, and Vlaic 1973), showed that Mn occupies only one of the two Ca positions, thereby establishing tinzenite as a valid species when that one position is more than half filled by Mn.

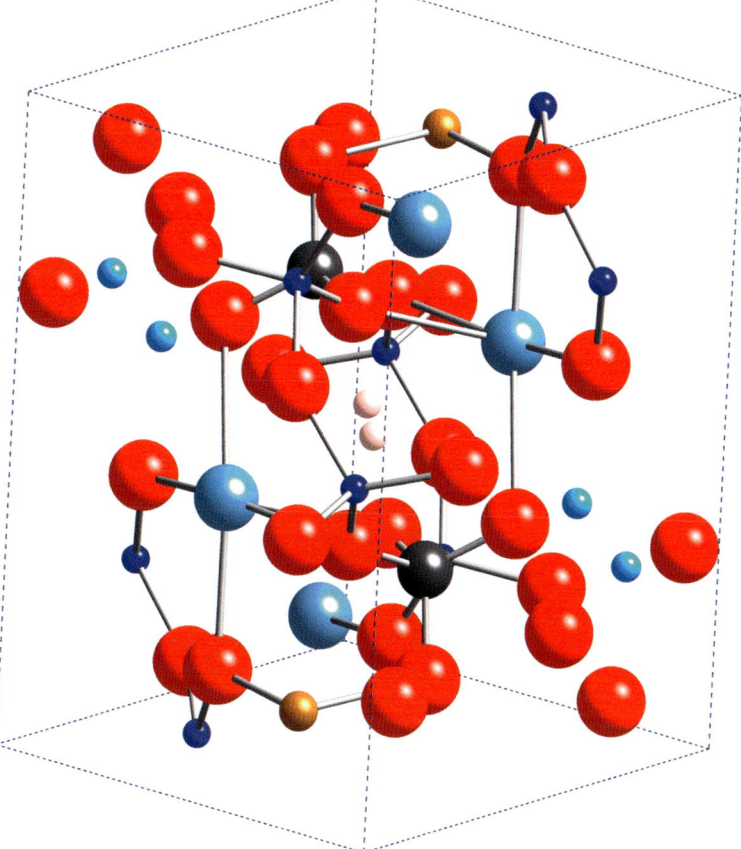

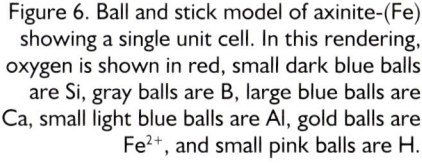

Figure 6. Ball and stick model of axinite-(Fe) showing a single unit cell. In this rendering, oxygen is shown in red, small dark blue balls are Si, gray balls are B, large blue balls are Ca, small light blue balls are Al, gold balls are Fe^{2+}, and small pink balls are H.

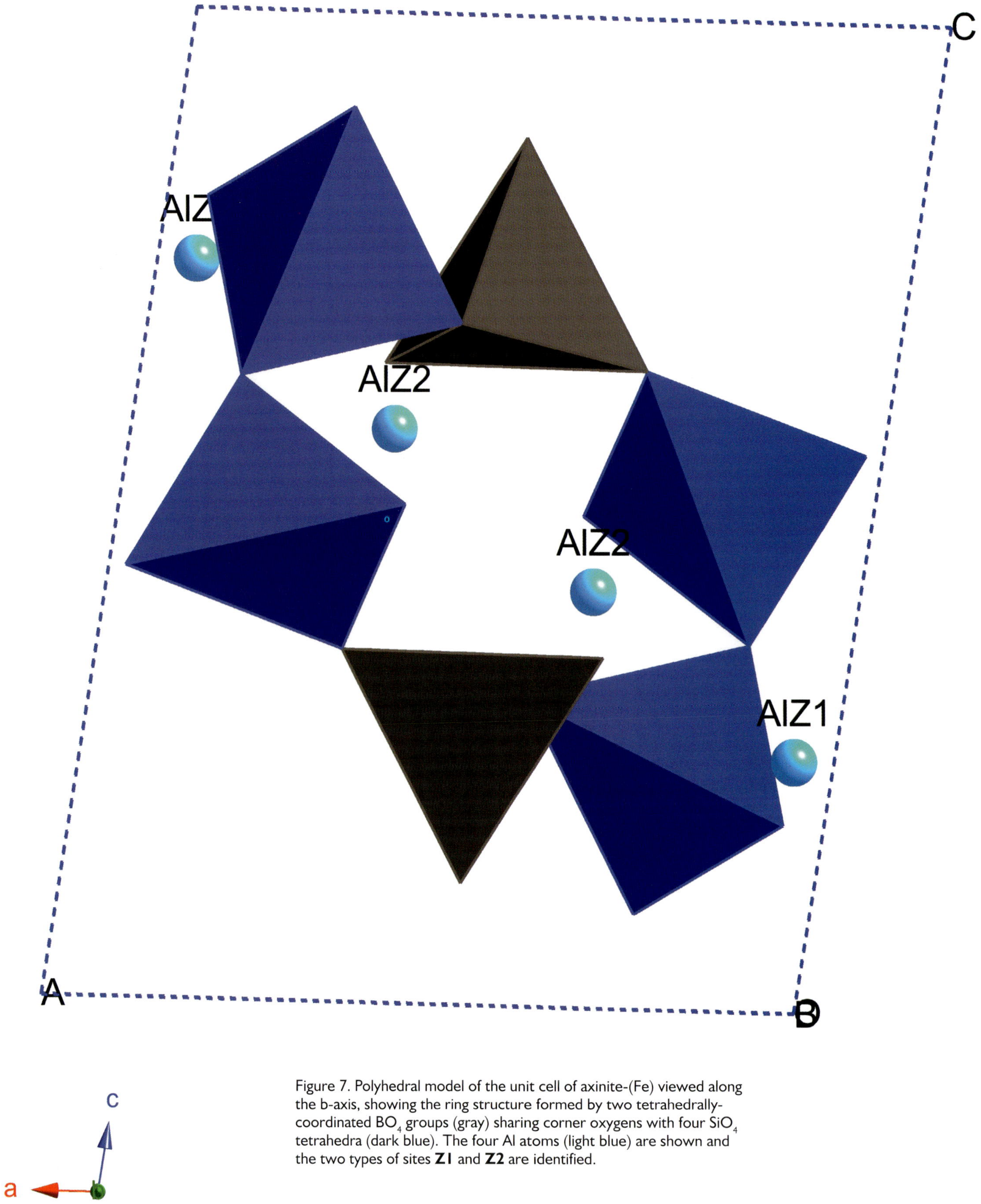

Figure 7. Polyhedral model of the unit cell of axinite-(Fe) viewed along the b-axis, showing the ring structure formed by two tetrahedrally-coordinated BO_4 groups (gray) sharing corner oxygens with four SiO_4 tetrahedra (dark blue). The four Al atoms (light blue) are shown and the two types of sites **Z1** and **Z2** are identified.

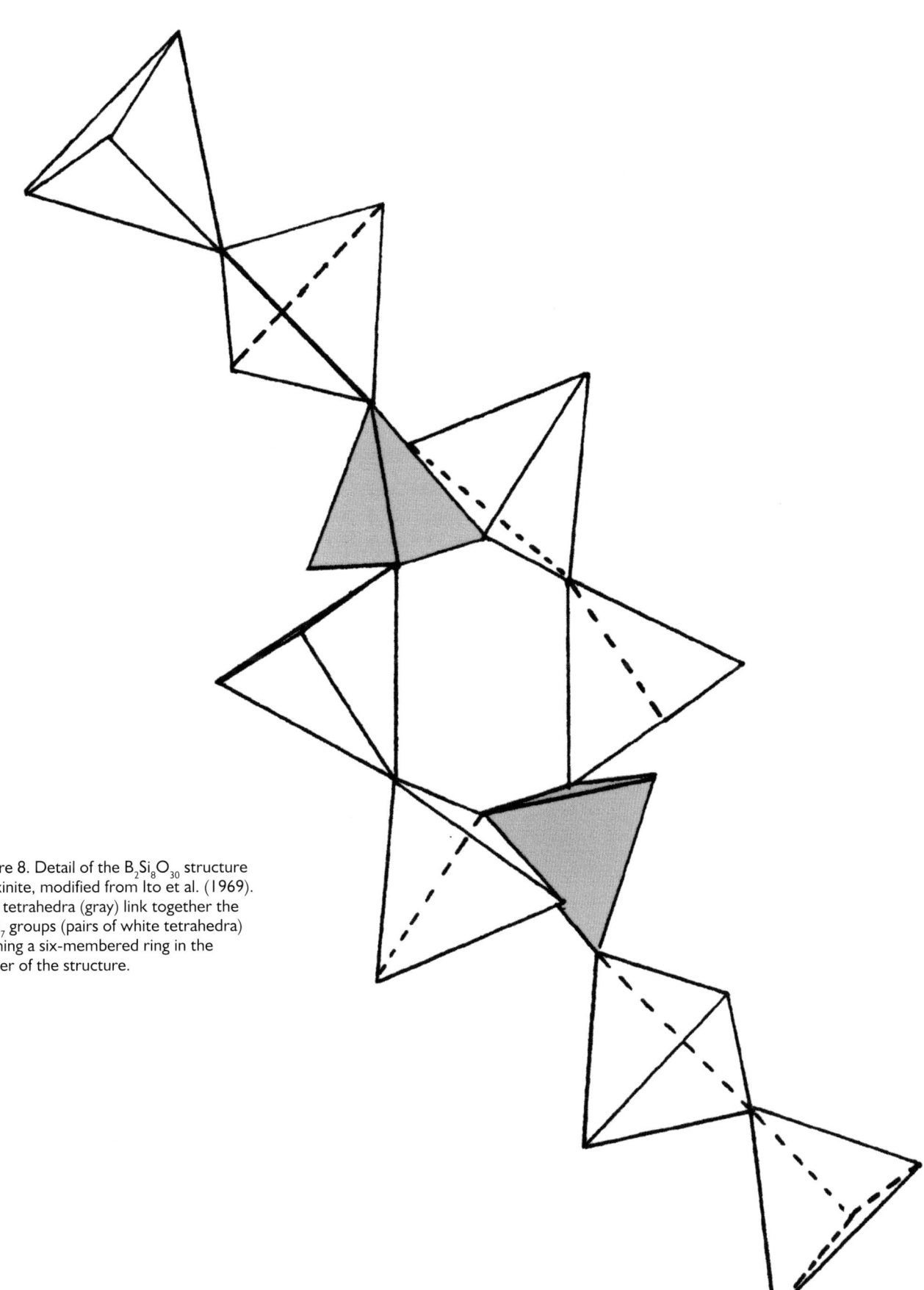

Figure 8. Detail of the $B_2Si_8O_{30}$ structure in axinite, modified from Ito et al. (1969). BO_4 tetrahedra (gray) link together the Si_2O_7 groups (pairs of white tetrahedra) forming a six-membered ring in the center of the structure.

Most axinites that the collector is likely to encounter have a nearly unmistakable habit: tabular or wedge-shaped axe-like crystals, often sharp enough to cut the fingers of a careless collector! Goldschmidt (1916) presented 172 drawings of natural crystals with their faces indexed to identify the crystallographic forms.

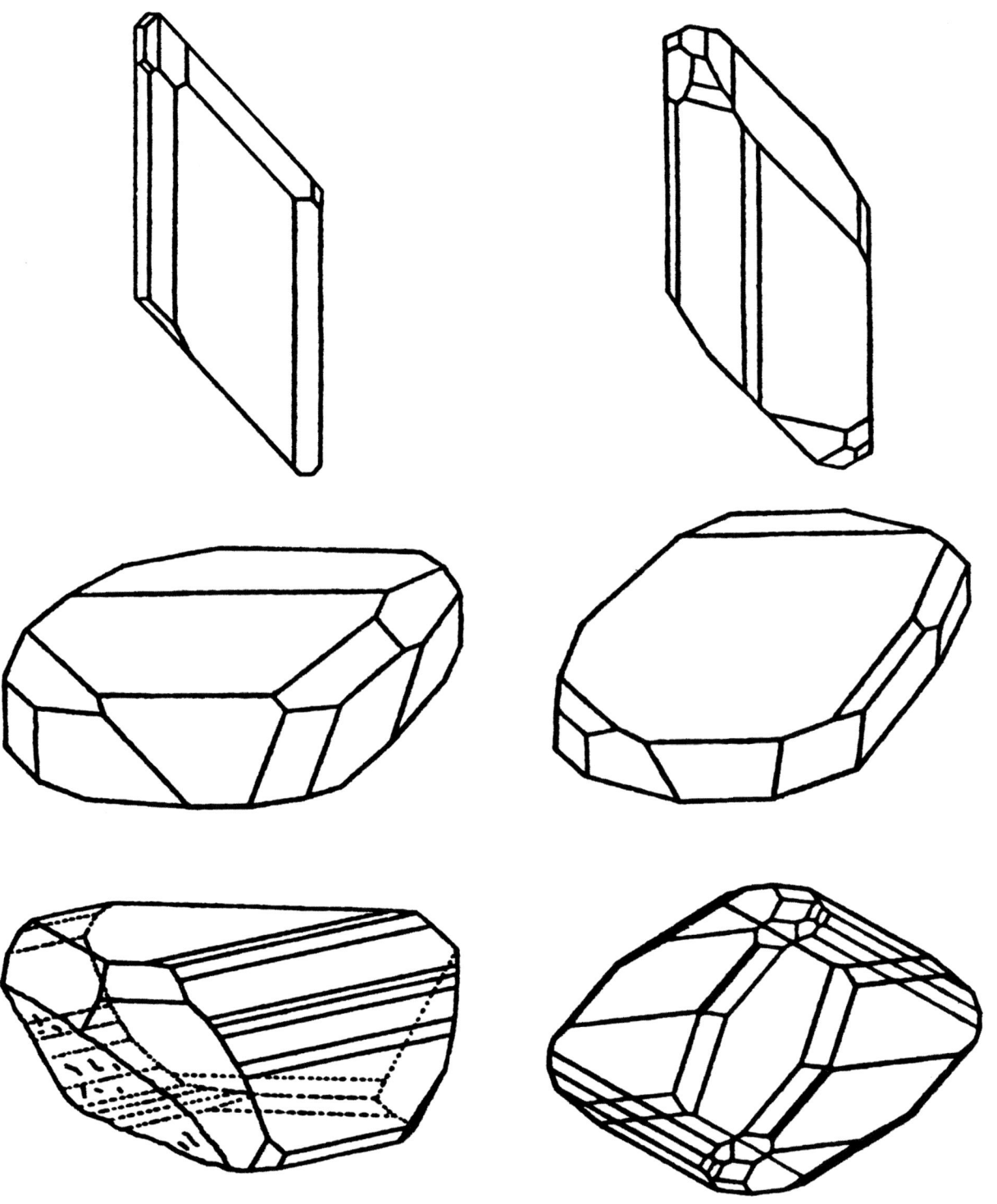

Figure 9. Drawings of natural axinite crystals modified from Goldschmidt (1913). Top: acute tabular crystals with sharp edges; center: tabular crystals without acute corners or bladed edges; bottom: equant crystals with striations representing a number of minor forms (faces).

Figure 10. Axinite from Dal'negorsk illustrating the acute morphology similar to those shown in the top row of Fig. 9. *RJL3276*

Figure 11. Axinite from Nandan, Guangxi province, China, illustrating the tabular morphology similar to those shown in the middle row of Fig. 9. *RJL3266*

Figure 12. Axinite from La Trinidad, Baja California, Mexico, showing a more equant habit. *RJL3273*

Formation and Geochemistry

The usual habitat for axinite is in contact metamorphic aureoles, typically where there has been metasomatic introduction of boron into existing calcareous rocks. The original source of boron is normally an acid magma, so a common setting would be in an impure limestone that has been metamorphosed by a nearby granite intrusion (Deer, Howie, and Zussman 1986). Within these metamorphic aureoles the axinite crystals are often found in veins filling contact zones, vugs, and tension fractures. More rarely, axinites are found in some pegmatites, skarns, and regionally metamorphosed rocks.

Axinites in Igneous Rocks

Occurrences of axinite where the mineral is unambiguously of igneous origin are fairly rare compared to metamorphic occurrences. In the Russian Mission quadrangle, Alaska, cm-sized clusters of small (1 – 2 mm) axinite-(Fe) crystals occur with larger twinned crystals of plagioclase and orthoclase embedded in a mixture of quartz and altered feldspars; the axinite is interpreted as a primary constituent of a quartz monzonite. According to the authors, "The occurrence of axinite in the quartz monzonite may be unique for the axinite crystallized as a primary constituent from a boron-bearing magma. Normally, boron-bearing solutions escape from the crystallizing magma, and axinite is deposited with quartz in veins or pegmatites cutting the host rock or in skarns in intruded carbonate rocks. Here, however, boron was probably trapped locally in the crystallizing rock and formed axinite in the presence of abundant calcium and sufficient aluminum and iron." (Hietanen and Erd 1978).

Axinites are occasionally found in pegmatites. Axinite-(Fe) has been noted from the Himalaya and Little Three pegmatites, San Diego County, California; and the Foote spodumene mine, Cleveland County, North Carolina. Axinite-(Mn) occurs at the Little Three pegmatite, California; the McKinney pegmatite, Spruce Pine, North Carolina; and the Leipers pegmatite, near Swarthmore, Pennsylvania (King 2003). Interestingly, axinites were rather rare in four of the five dikes of the Little Three pegmatite complex, and were abundant only in the Hatfield Creek dike (Stern et al. 1986).

Axinite-(Mn) is also found in a pegmatite vein cutting the Biwabik iron formation, Mesabi Range, Minnesota, associated with quartz, chlorite, and K-feldspar. The axinite did not show crystal faces, but rather formed grayish-brown patches of extensively fractured grains. Chemical analysis showed the crystals to be very Mn-rich ($\sim Mn_{0.8}Fe_{0.2}$). The authors concluded, "The presence of a highly manganoan axinite (manganaxinite) in a vein surrounded by metamorphosed iron-formation and associated with a gabbroic intrusion is noteworthy and suggests that the pegmatite veins containing the axinite were not derived in any significant amount from the iron-formation," (French and Fahey 1972).

A near-end-member axinite-(Fe) ($\sim Fe_{0.95}Mn_{0.05}$) was found in a highly Fe-contaminated granitic pegmatite dike cutting magnetite-rich sections of an Fe-rich skarn near Malešov, Czech Republic. The formation of such an Fe-rich axinite in the late stage of pegmatite evolution is attributed to the high activity of Fe and almost complete lack of Mg and Mn in the parent fluids. Furthermore, temperature plays a role, because higher temperatures favor more miscibility whereas lower temperatures favor compositions closer to end-member. At Malešov, the axinite-(Fe) is associated with prehnite and chlorite, indicating hydrothermal conditions well below 300°C during the late pegmatite evolution (Filip et al. 2006).

Another pegmatite occurrence is reported at Strzegom, Lower Silesia, Poland (Bernard and Hyršl 2004).

An unusual tin-bearing tinzenite has been reported from a niobium-yttrium-fluorine (NYF type) pegmatite from the Trebič pluton, Moldanubicum, Czech Republic. This occurrence represents the first reported occurrence of tinzenite in pegmatites and the second report of the incorporation of Sn into natural axinites (Škoda and Čopjakova 2006).

Figure 13. Two small, intergrown crystals of axinite-(Mn) from the Little Three pegmatite, San Diego County, California. *RJL2983*

Axinites in Metamorphic Rocks

For collectors, the finest axinite specimens are usually formed by contact metamorphism and metasomatism. Important occurrences include skarns and fissure or Alpine cleft-type deposits. Axinite occurrences in regional metamorphic deposits are known but are of somewhat less importance.

Skarn deposits include the pre-eminent axinite locale at Dal'negorsk in the Russian Far East, where superb crystals to 10 cm long are found in cavities, often associated with calcite and muscovite (Moroshkin and Frishman 2001). Excellent axinite-(Fe) crystals are also found in skarn at Pribrezhnoye, Chukotka, Russia, which was discovered in 1985 and prospected for the mineral specimen market in 1991. The crystals have similar morphology to the axinites from Dal'negorsk (Kalachev 1993). Axinite-(Fe) crystals have been known for nearly two hundred years from several mines in Cornwall, England. These classic locales include contact skarns, as well as calc-silicate hornfels attributed to alteration of volcanic rocks in contact with the Land's End granite. The Meldon quarry, Okehampton, Devon, was mined for railway ballast since the late 1800s (Embrey and Symes 1987). The occurrence is described as follows: A Na- and Li-rich aplite dike about 20 m thick occurs about 1 km northwest of the main Dartmoor granite; in the area of the Meldon quarry this dike forms contact skarns with a basic doleritic dike to the south, with calcareous shales to the northwest, and with "calc-flintas" (metasediments having alternating calcareous and siliceous bands) to the northeast. Although the samples were removed from rocks whose original compositions were very different, none of the axinites at Meldon showed significant chemical differences; in particular, there was no relation between the Fe, Mn, and Ca contents of the host rocks and the axinites. The chemical data suggest that the axinites were not formed through simple metamorphic reconstitution of the host rocks but rather by metasomatic introduction of large amounts of MnO, FeO, and B_2O_3 (Chaudhry and Howie 1969). At Colebrook Hill, Rosebery district, Tasmania, Australia, axinite-(Fe) occurs in a complex skarn containing sulfide ore minerals including pyrite, chalcopyrite, pyrrhotite, and arsenopyrite. An interesting occurrence of axinite-(Mg) containing approximately 0.13% Sn in skarns associated with tin-bearing granites has been described from eastern Siberia (Nekrasov 1971).

The first documented occurrence of axinite in Mexico was, interestingly, near Mapimi in the Ojuela mining district better known to collectors for superb secondary minerals such as adamite and legrandite. In the Vinagrillos hills about 8 km east of Mapimi, axinite-(Fe) is found as concretions with rounded to irregular shapes up to about 20 cm in diameter, or more rarely in bands and lenses to 5 cm thick. In the interior of the concretion zones, tiny (0.1 to 1.5 mm) crystals are found in small (2 to 4 mm) open spaces. The geological setting is described as follows: "The host rock is limestone or fine sandstone of marine origin ... (late Cretaceous). The area shows an early Tertiary intensive igneous activity with the main system consisting of dacite and latite intrusive bodies," (Swinnea et al. 1981).

Figure 14. Sharp bladed brown crystals of axinite-(Mn) associated with quartz, on axinite skarn from Dal'negorsk, Russia. Specimen is about 6 cm across. *RJL2537*

Figure 15. Closer view of the specimen in the previous figure, showing massive axinite in the skarn gradually becomes more sharply crystalline at the wall of the pocket where larger crystals ultimately grew.

Figure 16. Another view of the previous specimen showing the razor-sharp edges and "split" or "multiheaded" shape often seen at Dal'negorsk.

Figure 17. Brown crystals of axinite-(Fe) to 8 mm, lining cavity in skarn from the historic Meldon quarry, Okehampton, Devon, England. *RJL3269*

Figure 18. Purple-brown tabular axinite-(Fe) on green actinolite from Colebrook Hill, Tasmania, Australia. *RJL3290*

Fissure or cleft deposits include many important axinite locales. The Puiva mine, Subpolar Urals, Russia, is a famous Alpine cleft-type deposit that was mined intermittently for several decades, primarily for industrial quartz crystals. Beginning in the early 1990s, mining was directed to "collector" specimens and faceting rough, yielding spectacular axinite-(Fe) crystals that are among the world's finest examples of the species (Burlakov 1999). Sharp axinite-(Fe) crystals are also found in Alpine cleft-type deposits at Alchuri, Pakistan (Hammer and Weerth 2004). At New Melones Lake, California, axinite-(Fe) crystals to 10 cm are found in tension cracks in metabasalt and metagabbro in association with colorless quartz and occasionally epidote, chlorite, and actinolite (Pohl et al. 1982).

Figure 19. A sharp, 2 cm axinite-(Fe) crystal from Massif de Belledonne, Isere, France, showing a typical cleft-type mineral association. *RJL3268*

Figure 20. Typical axinite-(Fe) crystals to 3 cm scattered on matrix from Puiva, Subpolar Urals, Russia. *RJL2138*

Figure 21. A stunning, large plate of axinite-(Fe) crystals from Tomas, Pakistan. Specimen is about 15 cm across. *RJL-2408*

Figure 22. Bladed pinkish-brown axinite-(Fe) crystals in a cluster associated with colorless quartz, from New Melones Lake, California. *RJL-2538*

Formation and Geochemistry 39

Figure 23. Axinite-(Fe) from the Northern Areas, Pakistan, showing typical cleft association with a fibrous silicate (possibly actinolite). *RJL-3251*

Figure 24. Sharp, transparent brown axinite-(Fe) crystals to about 3 cm, with white orthoclase var. *adularia* and small green crystals (probably titanite) from an Alpine cleft-type deposit in Kohlistan, Pakistan. *RJL-3304*.

Regionally metamorphosed rocks occasionally host axinites. In southern New Zealand ferro-axinite, manganaxinite, and tinzenite are found in low-grade regionally metamorphosed rocks in areas where there is no evidence of igneous intrusions or contact metamorphic processes (Pringle and Kawachi 1980). At Baixa da Brauna, Bahia, Brazil, sharp axinite-(Fe) crystals are found in calcite veins cutting a sheared serpentinite overlain by garnetiferous amphibolite and underlain by quartzite. At this locale, two types of specimens are found: 1. partial crystals to about 10 cm and flattened aggregates on a hornblende matrix; and, 2. small crystals disseminated in dense white calcite. Although the site produced some fine crystals, it is perhaps not as well known to collectors as it should be, simply because mining activities were only conducted for a short time by miners who apparently mistook the hornblende for tourmaline (Cassedanne and Cassedanne 1977).

Figure 25. A sharp, purple-brown striated crystal of axinite-(Fe), about 5 cm long, from Baixa da Brauna, near Vitoria da Conquista, Bahia, Brazil. *RJL3272*

The Minerals

When axinite crystals were first discovered around 1780 in France at Bourg d'Oisons, they were thought to be a variety of tourmaline. During the next twenty years, various authors described material from several localities under a variety of names. Haüy first used the modern name "axinite" in 1799, in allusion to the sharp, axe-like crystals.

Axinite-(Fe)

Axinite-(Fe) was recognized as a distinct species in 1909 and historically called ferro-axinite as the Fe-dominant member of the group. It may be grayish, blue-gray, various shades of brown, or greenish. Large crystals are usually dark brown but transparent at thin edges. Axinite-(Fe) is widely distributed and occurs in a number of settings, including metamorphic/metasomatic zones, alpine-type veins, quartz-calcite veins, and occasionally in pegmatites. Many localities produce excellent crystals and interesting associations, making it difficult to single out a few "classic" occurrences that everyone would agree on.

Bourg d'Oisans, Isere, France, has long been known for sharp, transparent brown crystals that were, perhaps, the original standard of quality and crystal habit. Elsewhere in Europe, good examples occur in many alpine deposits in Germany, Switzerland, Czech Republic, and Austria. In Italy, crystals to 7 cm have been reported from Val d'Ossola, near Premia. In Britain, dark brown crystals were found at the Botallack mine in Cornwall, and pink to tan crystals have been recovered from the seaside bluffs at Stamps and Jowl Zawn, Roscommon Cliff. The Cornish localities were known in the early 1820s (Embrey and Symes 1987).

Figure 26. Sharp brown axinite-(Fe) crystals to about 2 cm, forming a plate with minor matrix, from Bourg d'Oisans, Dauphine, France. RJL1291

42   The Minerals: Axinite-(Fe)

Figure 27. An example of purple-brown axinite-(Fe) crystals illustrating the cleft-type occurrence at Bourg d'Oisans, France. *RJL3271*

The Minerals: Axinite-(Fe) 43

Figure 28. A thumbnail-sized cluster of axinite-(Fe) crystals with albite, from Piz Vallatscha, Lukmanier Pass, Grisons, Switzerland. Specimen dates from the 1920s. *RJL3018*

44　The Minerals: Axinite-(Fe)

Figure 29. Sharp brown crystals to about 1 cm thickly covering the surface of cleft-type matrix, from Scopi, Switzerland. *RJL3363*

Figure 30. Brown crystalline to massive axinite with minor chlorite, from St. Maria, Switzerland. *RJL3371*

Figure 31. Small brown bladed axinite crystals with minor epidote, from Cantera Juanona, Antequera, Malaga, Spain. *RJL3260*

The Minerals: Axinite-(Fe) 47

Figure 32. Pale brown axinite crystals to about 3 mm, with prehnite and quartz, from Bustarviejo, Madrid, Spain. *RJL3267*

Figure 33. An older specimen from a classic locale: Botallack mine, St. Just, Cornwall, England; brown crystals scattered on and partly comprising matrix. *RJL2420*

Several localities in Japan have yielded fine, dark brown axinite-(Fe) specimens. The Obira mine, Bungo, Oita Pref., and the Toraku mine, Iwate, Miyazaki Pref., are noteworthy examples.

Figure 34. Dark brown crystals in divergent clusters without matrix, from the Obira mine, Japan. Specimen is about 6 cm wide. *RJL2606.*

Figure 35. Dark brown axinite-(Fe) crystals in a parallel aggregate, about 3 cm tall, that looks like a crude single crystal, from the Toroku mine, Japan. *RJL1298*

In recent years, Russia has produced outstanding specimens from the Puiva deposit in the Tyumen district, Subpolar Urals, where it is common in many crystal pockets and clefts. One large cleft yielded over 200 kg of high-quality axinite-(Fe) specimens, some with crystals exceeding 20 cm! The crystals are purple-brown, transparent, and sharp-edged. Associated species include actinolite, calcite, chlorite, clinozoisite, datolite, fluorite, orthoclase var. *adularia*, and titanite (Burlakov 1999). [As with many Russian locales, collectors can expect to see various transliterations of the name, such as Puyva, Pripolar Urals, and other variants.]

Figure 36. Sharp brown axinite-(Fe) crystals forming a 3-cm group on white orthoclase var. *adularia*, from Puiva, Subpolar Urals, Russia. *RJL3303*

The Minerals: Axinite-(Fe) 51

Figure 37. Purplish brown bladed crystals thickly standing on edge, cemented together by fine green chlorite, from Puiva, Russia. *RJL2179*

Figure 38. Rear of specimen in the previous figure, showing a thick layer of green chlorite, a mineral often seen in cleft-type deposits.

52   The Minerals: Axinite-(Fe)

Figure 39. A razor-sharp single crystal, heavily included with chlorite, giving it a dull green color. This crystal, also from the Puiva deposit, is about 4 cm wide but only about 6 mm thick. *RJL3360*

Figure 40. Dark brown intergrown axinite-(Fe) crystals, forming a group about 4 cm wide without matrix, from Chukotka, Russia. *RJL3355*

Figure 41. Lustrous, dark brown crystals to about 1 cm wide, in stacked groups on matrix. This old specimen of axinite-(Fe), from Ekaterinburg, Urals, Russia, shows an interesting habit: tabular crystals with rounded corners and edges. *RJL3356*

Comparable axinite-(Fe) specimens have been found in cleft type deposits near Tomas and elsewhere in the Northern Areas, Pakistan.

Figure 42. Extremely sharp bladed crystals standing up on matrix in a distinctive habit from a less-common locale in the Kharan district, Baluchistan, Pakistan. The largest crystal is 6 cm long. *RJL3186*

The recent flow of mineral specimens from China has included many fine examples of axinite, frequently associated with other interesting species. Collectors will notice that some pieces may be labeled ferro-, or mangan-, or in some cases simply "axinite." As always, in the absence of reliable published analyses, collectors must content themselves with the documentation accompanying the sample. This caveat applies to locality information as well. It is likely that as more material becomes available for study, and outside experts have greater opportunities to inspect the deposits firsthand, our knowledge will improve significantly, as it did a decade ago for localities in the Former Soviet Union. The Chinese specimens illustrated here are identified as to both species and locale in reliance on their accompanying labels; no attempt has been made to perform destructive or nondestructive analyses on them.

Figure 43. Typical brown axinite-(Fe) crystals to about 1 cm, thickly covering matrix associated with several colorless euhedral calcite crystals, from Nandan, Guangxi, China. *RJL2968*

Figure 44. Tabular axinite-(Fe) associated with elongated quartz crystals that are colored green by inclusions thought to be actinolite, From Wenshan, Yunnan, China. Specimen is about 6 cm tall. *RJL3302*

Figure 45. Detail of specimen in the previous figure showing the tabular morphology of the axinite.

The premier Australian locale for axinite-(Fe) is Colebrook Hill, Rosebery district, Tasmania. The deposit was originally worked for copper, but other base metals and precious metals have been found there. Well-crystallized specimens are generally found encased in massive crystalline calcite, which can be dissolved with acid to reveal fine axinite-(Fe) crystals. Associated minerals include fibrous green amphiboles of the tremolite-actinolite series, quartz, and various sufides including chalcopyrite, loellingite, arsenopyrite, and others.

Figure 46. Small purplish brown axinite-(Fe) crystals with massive sulfides, from Colebrook Hill, Tasmania, Australia. *RJL2675*

Figure 47. An association typical of material from Colebrook Hill, Australia: purple-brown axinite crystals to about 1 cm, with quartz and massive/fibrous tremolite-actinolite. *RJL2839*

Good crystals have been collected from several places in California, in a belt running generally NNW from Coarsegold in Madera County, through New Melones Lake in Calaveras County, to the Feather River area in Plumas County. In the Feather River area, sharp axinite crystals up to 2 cm long occur with white quartz in veins 3 to 15 cm thick cutting fine-grained metavolcanics (Hietanen and Erd 1978). A particularly important find occurred in the early 1970s during excavation of the spillway for the New Melones Lake Dam. There, veins filled with massive quartz and axinite-(Fe) follow tension fractures in basic dikes cutting metamorphosed basalts and gabbros. Where the veins expand to form open pockets, large and very lustrous crystals were collected both as individuals and as matrix specimens with quartz, epidote, actinolite, and albite. New Melones Lake is arguably the finest American occurrence for axinite-(Fe) (Pohl, et al. 1982).

Figure 48. A small transparent axinite-(Fe) crystal about 10 X 15 mm, from Yankee Hill, Feather River Canyon, Plumas County, California. *RJL3274*

Figure 49. Unusually large, sharp axinite-(Fe) crystals with minor epidote, from New Melones Dam, Calaveras County, California. Specimen is about 8 cm wide. *RJL3264*

Figure 50. Purple-brown blades of axinite-(Fe) with colorless, terminated quartz, from New Melones Dam, Calaveras County, California. Specimen is about 5 cm wide. *RJL3265*

The Minerals: Axinite-(Fe) 61

Figure 51. Purple-brown terminated crystals forming a flattened group about 6 cm long, without matrix, from New Melones Dam, Calaveras County, California. *RJL2300*

Figure 52. A small, sharp axinite-(Fe), possibly colored by unidentified inclusions, associated with heavily included quartz crystals, from Snohomish County, Washington. *RJL3253*

Axinite-(Fe) occurs at a number of localities scattered along a band that runs roughly from Bethlehem, Pennsylvania, to Fishkill, New York. (Interestingly, the classic mines near Franklin, New Jersey, lie near the center of this belt.) At the Oxford quarry, near Bridgeville, New Jersey, axinite-(Fe) is found in open fissures, typically on epidote. Other associated species include actinolite var. *byssolite*, albite, calcite, quartz, pyrite, and various manganese oxides. Petrographic analysis suggested that the late mineralization is the result of recrystallization of amphibolite rock by B-rich hydrothermal fluids, a genesis similar to that of the Alpine cleft deposits in Switzerland (Cummings 1983).

Figure 53. Purple-brown axinite-(Fe) crystals thickly filling a narrow fissure, associated with green epidote, from the Oxford quarry, Bridgeville, New Jersey. *RJL3357*

The Mayo mining district, Yukon Territory, Canada, is home to a number of silver-lead-zinc mines, many of which are closed, as well as tungsten-tin-molybdenum deposits. However, at one site, the Grey Cloud claim, interesting specimens of axinite-(Fe) with prehnite have been recovered.

Figure 54. Small, grayish-brown axinite-(Fe) crystals to about 6 mm associated with colorless to pale green prehnite crystals, from the Grey Cloud claim, Hart River, Mayo Mining district, Yukon Territory, Canada. *RJL3250*

In the Ica region, southern Peru, good crystals are found at several localities in association with other species, forming interesting and attractive specimens. At Molletambo, 40 km ESE of Ica, axinite forms crystals and groups to about 7 cm, occasionally with prehnite, black tourmaline, and epidote. Similar material is found at the Paracas quarry, near Huaytara, on the border of Ica and Huancavelica departments. Although the epidotes from these locales are not as spectacular as those from Pampa Blanca, the combination with axinite makes them unusual and interesting (Hyršl and Rosales 2003).

Figure 55. Sharp, purple-brown axinite-(Fe) crystals to about 2 cm on massive axinite, from Molletambo, near Ica, Peru. *RJL3147*

Figure 56. Gray-brown bladed axinite-(Fe) crystals associated with acicular green epidote in sheaflike aggregates, from the Castrovirreyna district, Peru. *RJL2193*

Axinite-(Mg)

As noted earlier, the original description of axinite-(Mg) is based on a faceted gemstone, now in the collection of the Natural History Museum – London (Jobbins, Tresham, and Young 1975). Specimens appear from time to time in the gem trade, but the species remains fairly scarce. At the type locality in the Arusha district, Tanzania, loose crystals found in the alluvial gem deposits are pale blue to blue-gray and their composition is very close to the ideal formula. The crystals tend to be rough and flattened, generally not as sharp and well formed as axinites from many localities. The faceted stone illustrated here has a noticeable color-change effect: blue-gray in fluorescent or daylight, lilac-purple under incandescent light. Furthermore, the stone fluoresces deep red in shortwave ultraviolet light. The type material contains a small amount of vanadium ($\sim$ 0.13 % V_2O_3), which might be responsible for the color effects.

Figure 57. A small faceted axinite-(Mg) (0.39 carat) photographed in tungsten (halogen) light shows a pale pink coloration.

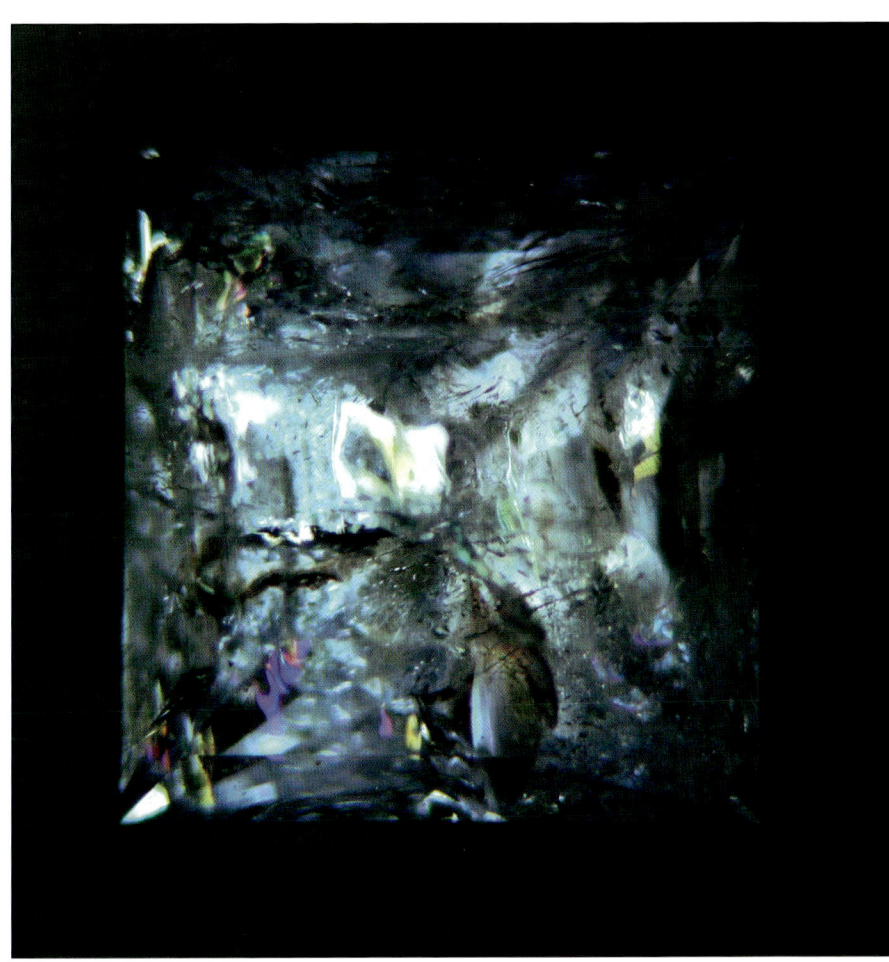

Figure 58. The same stone photographed in fluorescent light appears blue-gray.

Figure 59. Under SW UV, traces of dull red fluorescence can be seen.

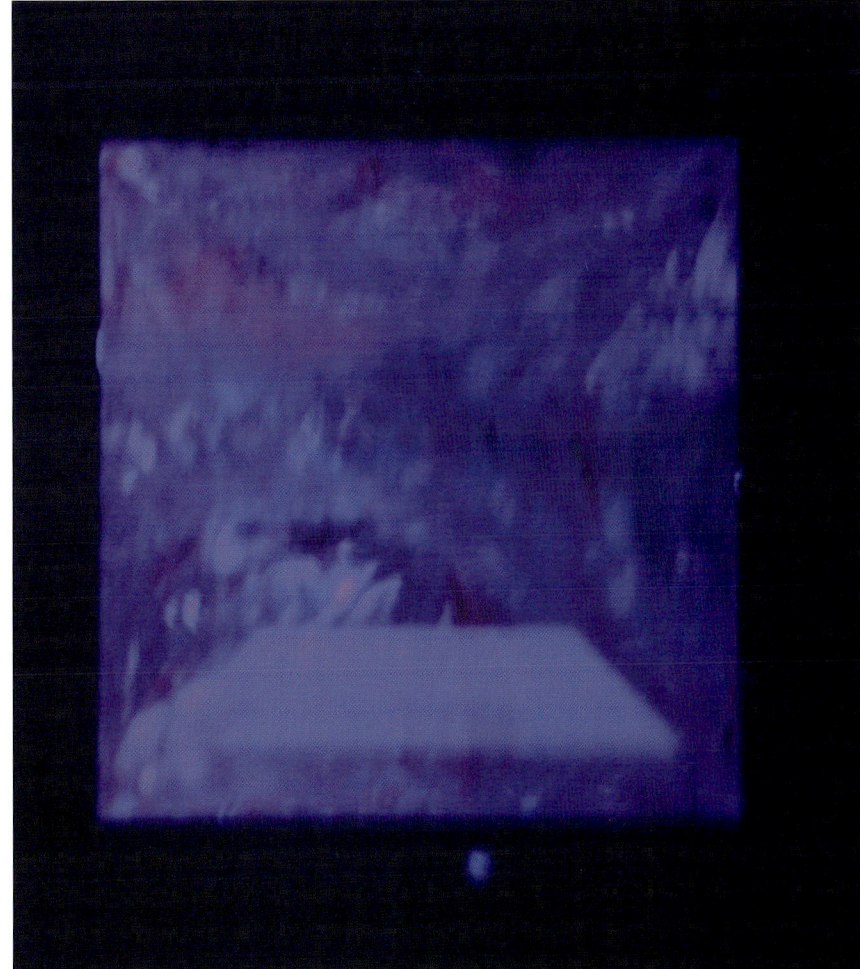

68   The Minerals: Axinite-(Mg)

Figure 60. A rough axinite-(Mg) crystal about 2 cm tall, from the Merelani Hills, near Arusha, Tanzania. *RJL3148*

Figure 61. Side view of specimen shown in the previous figure, illustrating the wedge-shaped tabular habit typical of the axinite group.

Ferroan varieties of axinite-(Mg) have been identified at several localities. At London Bridge, Queanbeyan, New South Wales, Australia, the crystals are found in a band cutting calcareous hornfels near quartz-feldspar porphyry, associated with epidote, tremolite, and prehnite. Their composition is very close to the axinite-(Fe) – axinite-(Mg) boundary but, with Mg:Fe around 1.2:0.8, they clearly lie within the axinite-(Mg) range (Deer, Howie, and Zussman 1986). At a site near Luning, Nevada, light brown to light pink crystals to about 1 cm are associated with prehnite, epidote, actinolite, and vesuvianite. Analyses showed that most of the crystals were axinite-(Mg) (containing significant Fe and Mn), while other samples were actually axinite-(Fe). The authors of that study suggest that an axinite crystal from this locality that has not been analyzed should be labeled axinite-(Mg), based on the preponderance of this species among the samples they studied (Dunn, Leavens, and Barnes 1980). Ferroan axinite-(Mg) is also found in hydrothermal veins at Lažany, Brno Batholith, Czech Republic, associated with clinozoisite and actinolite. The pale gray to pinkish-gray axinite is typically massive but occasionally forms small tabular crystals to about 5 mm in open vugs. As with the material from Luning, Nevada, some of the crystals from Lažany are magnesian axinite-(Fe) (Novak and Filip 2002).

Axinite-(Mg) was recently reported from the Bakal deposit, South Urals, Russia, where it occurs in dolomite in the contact zones adjacent to diabase dikes. The crystals exhibit typical axinite morphology and some compositional zoning in which the rim corresponds to $Mg_{0.46}Fe_{0.29}Mn_{0.22}$ whereas the core Mg concentration is $Mg_{0.38}Fe_{0.30}Mn_{0.29}$. The species has also been found in northern Kazakhstan.

Axinite-(Mg) containing 0.78 to 0.13% Sn in calcareous skarns associated with tin-bearing granites has been described from eastern Siberia. The average composition corresponds to $Mg_{0.60}Fe_{0.36}Mn_{0.19}$ and it was noted that samples with higher Fe are also richest in Sn. It was assumed that the tin is present as Sn^{4+} and that a substitution for Fe^{3+} of the form $4Fe^{3+} \rightarrow 3Sn^{4+}$ is more likely than substitution of Sn^{4+} for Si^{4+} on the silicate tetrahedron (Nekrasov 1971). This would seem to be a reasonable hypothesis, given that the ionic radius of Sn^{4+} is 0.71 Å, which is much larger than that of Si^{4+} (0.42 Å) but only slightly larger than that of Fe^{3+} (0.64 Å). However, later experimental syntheses of more Sn-rich material under reducing conditions by Nekrasov and Kashirtseva [summarized in Deer, Howie, and Zussman (1986)] suggested that the tin is present as Sn^{2+} (ionic radius 0.93 Å) but is partially replacing Si^{4+}.

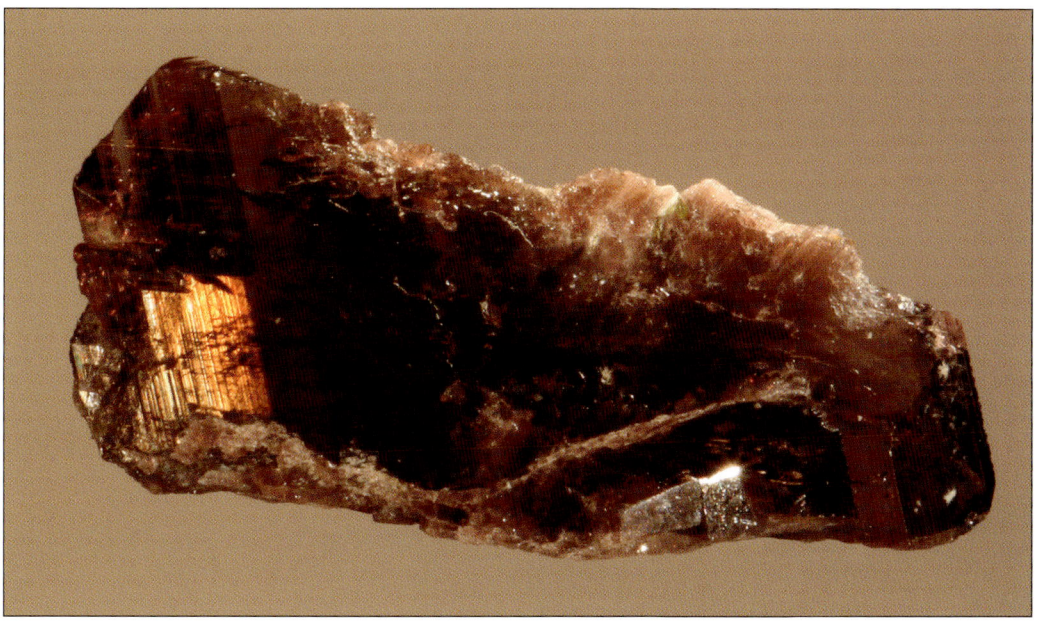

Figure 62. Transparent brownish axinite crystal from a locale near Luning, Nevada. As discussed in the text, the average composition at this locale lies very close to the boundary between axinite-(Mg) and axinite-(Fe), but the majority of analyzed samples fell within the axinite-(Mg) range. Thus, samples from this deposit that have not been analyzed should be presumed to be ferroan axinite-(Mg). RJL2626

The Minerals: Axinite-(Mg)

Figure 63. A pair of crude, pinkish-gray crystals of axinite-(Mg) from the Crestmore quarry, Riverside, California. Specimen is about 4 cm tall. *RJL3353*

Axinite-(Mn)

Axinite-(Mn) was known to occur at Franklin, New Jersey, in the late 1800s, and the name manganaxinite was applied in 1909. The Franklin material may be pale yellow to white massive grains or colorless to dark yellow euhedral crystals. It typically contains up to 5 percent Zn in substitution for Mn. Some of the paler crystals fluoresce bright red under both short- and long-wave ultraviolet light (Dunn 1979).

Figure 64. A sample about 15 mm tall consisting of pale cream-yellow axinite-(Mn) on massive brown andradite, from the Parker Shaft, Franklin, New Jersey. *RJL3373*

Figure 65. Small glassy pale yellow crystals of axinite-(Mn) associated with colorless quartz and brown hendricksite from Franklin, New Jersey. *RJL3372*

Sharp brown axinite-(Mn) crystals have been found in pockets in several dikes at the Little Three mine, San Diego Co., California (Foord et al. 1989; Stern et al. 1986). Although not as spectacular as the Russian material, they are perhaps the finest example of axinite from a pegmatite in the continental U.S.

Figure 66. A tabular single crystal of axinite-(Mn) about 20 X 25 mm, from the Little Three pegmatite, San Diego County, California. *RJL3263*

The Dal'negorsk mining district in the Russian Far East has produced superb specimens from both the lead-zinc deposits and the borosilicate deposits. In some areas, the skarns may contain as much as 90 percent axinite, most of which is massive. In open pockets and veins, the axinite-(Mn) forms sharp, purple-brown crystals up to 10 cm long, with clear quartz and epidote. The crystals are described as "usually split, having multi-headed shape and lustrous facets" (Moroshkin and Frishman 2001; see also Grant and Wilson 2001). Axinite crystal groups associated with other species, including quartz, fluorite, calcite, and datolite, make attractive and interesting specimens. Axinite-(Mn) pseudomorphs after hedenbergite are also occasionally found at Dal'negorsk (Lisitsyn and Malinko 1994).

Figure 67. Sharp group of deep brown, slightly curved axinite-(Mn) crystals from the Sentyabrskiy mine, Dal'negorsk, Russia. Largest crystal is about 6 cm tall. *RJL2139*

Figure 68. Lustrous brown axinite-(Mn) crystals completely covering matrix, from the Verchniy mine, Dal'negorsk, Russia. Tiny doubly-terminated quartz crystals are sprinkled over the axinite. *RJL2877*

76 The Minerals: Axinite-(Mn)

Figure 69. Another specimen from the Verchniy mine showing the same association as in the previous figure: divergent groups of axinite crystals dusted with colorless quartz. *RJL3289*

Figure 70. Lustrous axinite-(Mn) crystals in divergent groups completely covering a thin matrix plate, from the Verchniy mine, Dal'negorsk, Russia. *RJL2354*

Figure 71. An interesting association from the Verchniy mine, Dal'negorsk: purple-brown axinites to about 3 cm tall, thickly covering matrix and partially covered by tiny pinkish calcite crystals. The calcite is fluorescent (reddish) in SW UV. *RJL2859*

Figure 72. Brown axinite-(Mn) crystals and groups to about 1 cm tall, associated with a 3-cm datolite crystal, from the Bor Pit, Dal'negorsk. *RJL3275*

Figure 73. A complex, nearly spherical colorless fluorite crystal about 1 cm in diameter on small brown blades of axinite-(Mn), from the Verchniy mine, Dal'negorsk. *RJL3259*

Figure 74. Two hexagonal "poker chip" calcite crystals, the larger about 4 cm across, standing on a thin plate of axinite-(Mn), from the Verchniy mine, Dal'negorsk. *RJL2396*

82 The Minerals: Axinite-(Mn)

Figure 75. A striking association: a colorless 4 cm long, terminated quartz crystal on a lustrous axinite 7 cm tall, from the Bor Pit, Dal', Dal'negorsk. RJL3383

Figure 76. Side view of specimen in the previous figure, nicely illustrating the curved habit of the axinite.

Some other reported axinite-(Mn) localities include: the McKinney mine, Mitchell County, North Carolina; in the Huachuca Mts., Arizona; Avondale, Pennsylvania; the Mitchell mine, Babbitt, Minnesota; Marmora Township, Ontario, Canada; Monte Pu, Liguria, Italy; in Japan at the Anawai mine, Shikuko, and at the Takanosu mine, Tachigi Pref.; at Nandan, Dachang tin area, Guangxi, China; and in the Julcani district, Peru.

Figure 77. Small greenish tabular axinite-(Mn) crystals thickly lining vugs in matrix, from the Iron Cap mine, Landsman Camp, Graham County, Arizona. *RJL3354*

Figure 78. A closer view of the crystals in the previous photo. Field of view is about 8 mm.

Figure 79. Lustrous purplish brown axinite crystals on matrix, associated with green fluorite crystals, from Nandan, Dachang tin area, Guangxi, China. *RJL3161*

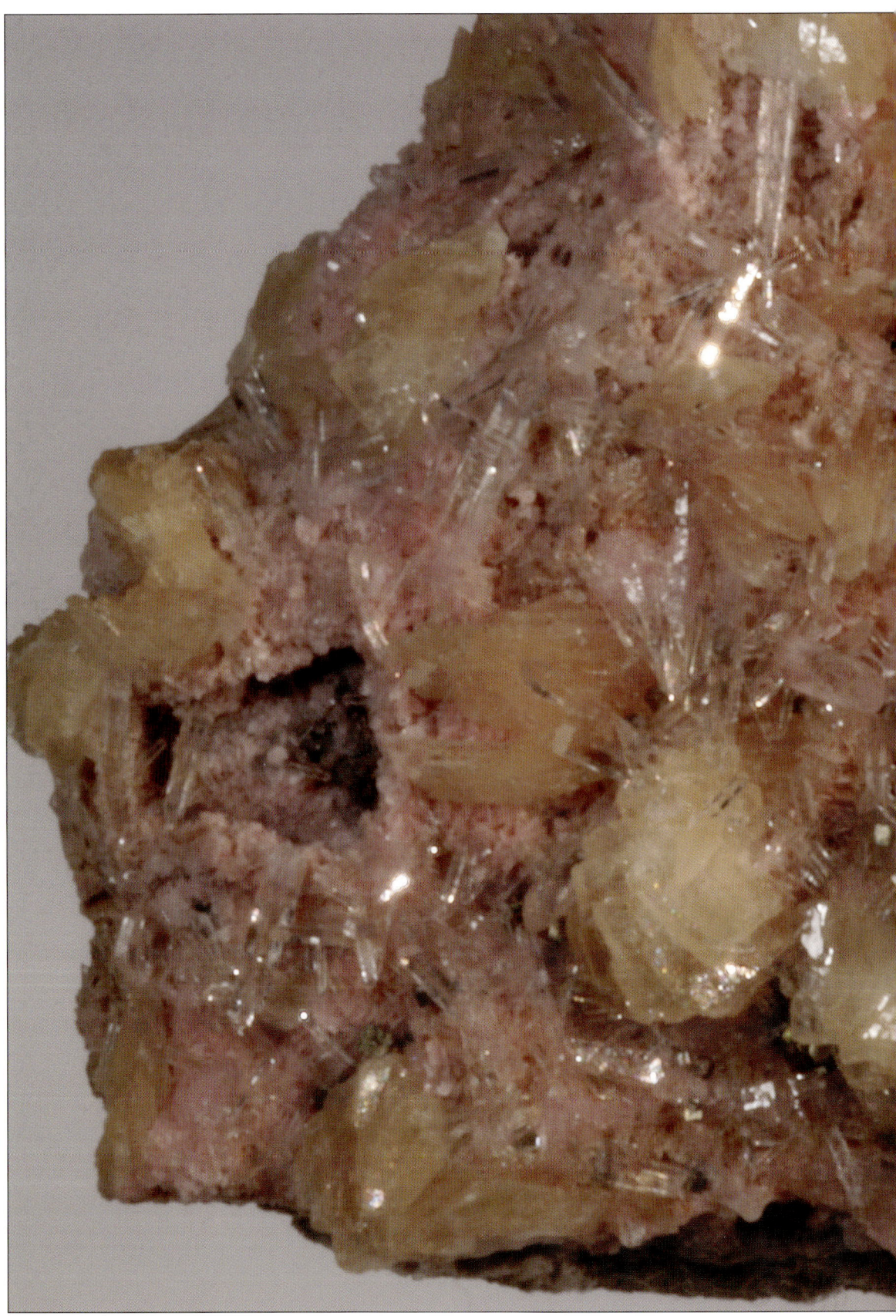

Figure 80. Pale yellow bladed crystals of axinite-(Mn) to about 3mm, in clusters on matrix of rhodonite, quartz, and pyrite, from the Pachapaqui district, Ancash, Peru. *RJL3171*

Tinzenite

Tinzenite was originally described in a German paper (Jakob 1923) from a site near Tinzen, in the Val d'Err, Graubünden (= Grisons or Grischun), Switzerland. Detailed chemical analysis and X-ray studies (Milton, Hildebrand, and Sherwood 1953) showed that it was isostructural with members of the axinite group. Its crystals tend to be somewhat smaller than those of the other axinites, typically 5 mm, but they show the typical wedge-shaped or bladed habit. The color ranges from pale yellow to various shades of orange and pink.

Tinzenite is not as widely distributed as axinite-(Fe) or axinite-(Mn), and occurs mainly in metamorphosed manganiferous sediments in the Tinzen region of Switzerland and in Liguria, Italy. The Italian deposits are noteworthy for both the quality of the tinzenite and the extraordinary variety of other rare minerals found in the district (Marchesini and Pagano 2001): In the Gambatesa mine, yellow to rose-colored tinzenite is associated with yellowish-green or brown manganaxinite, piemontite, and spessartine in sedimentary manganese ore (Deer, Howie, and Zussman 1986). At the Molinello mine, bright orange bladed aggregates with clear quartz make attractive specimens. It also occurs at the Monte Bossea mine. Tinzenite is also reported from low-grade metamorphic rocks in New Zealand, but generally, axinite-(Mn) tends to be more prevalent than tinzenite in most of the localities there. Small yellow crystals associated with rhodonite have been reported from the Pachapaqui mine, Ancash, Peru.

An unusual variety of tinzenite containing up to 0.86% SnO_2 has been reported from a niobium-yttrium-fluorine (NYF type) pegmatite from the Trebič pluton, Moldanubicum, Czech Republic. This occurrence represents the first reported occurrence of tinzenite in pegmatites and the first report of the incorporation of Sn into natural axinites (Škoda and Čopjakova 2006).

Figure 81. A thumbnail-sized sample of massive to bladed orange tinzenite in metamorphosed manganiferous rock, from the type locale, Falotta, Tinzen, Grisons, Switzerland. *RJL3254*

Figure 82. Pinkish-orange tinzenite as subparallel groups of thin bladed crystals with colorless quartz from the Molinello mine, Italy. Sample is about 3 cm across. *RJL2623*

90 The Minerals: Tinzenite

Figure 83. Tiny yellow, bladed, 3-5 mm tinzenite crystals on drusy pink rhodonite from the Pachapaqui mine, Ancash, Peru. *RJL2922*

References

Andreozzi, G.B., L. Ottolini, S. Lucchesi, G. Graziani, and U. Russo 2000. "Crystal chemistry of the axinite-group minerals: A multi-analytical approach." *American Mineralogist* 85:698-706.

Andreozzi, G. B., S. Lucchesi, G. Graziani, and U. Russo 2004. "Site distribution of Fe^{2+} and Fe^{3+} in the axinite mineral group: New crystal chemical formula." *American Mineralogist* 89:1763-71.

Anthony, J. W., S. A. Williams, and R. A. Bideaux 1982. *Mineralogy of Arizona*. Tucson: The University of Arizona Press.

Arem, J. 1977. *Color encyclopedia of gemstones*. New York: Van Nostrand Reinhold.

Basso, R., A. Della Guista, and G. Vlaic 1973. "La struttura della tinzenite," *Periodico Min.* 42: 369-79.

Bauer, M. 1904. *Precious Stones*, (1968 reprint) New York: Dover Publications, 627 pp.

Bernard, J., and J. Hyršl. 2004. *Minerals and their localities*. Prague: Granit.

Burke, E. A. J. 2008. "Tidying up mineral names: an IMA-CNMNC scheme for suffixes, hyphens and diacritical marks," *Mineralogical Record* 39 (2): 131-5.

Burlakov, E. V. 1999. "The Puiva deposit, Subpolar Urals, Russia," *Mineralogical Record* 30 (6): 451-65.

Cassedanne, J. P. and J. O. Cassedanne 1977. "Axinite, hydromagnesite, amethyst and other minerals from near Vitoria da Conquista (Brazil)," *Mineralogical Record* 8 (5): 382-7.

Chaudhry, M. N. and R. A. Howie 1969. "Axinites from the contact skarns of the Meldon aplite, Devonshire, England," *Mineralogical Magazine* 37 (285): 45-8.

Crowley, J. A., R. H. Currier, and T. Szenics 1997. "The mines and minerals of Peru," *Mineralogical Record* 28 (4):7-98.

Cummings, W. 1983. "Ferro-axinite from Bridgeville, New Jersey," *Mineralogical Record* 14 (1): 43-4.

Deer, W. A., R. A. Howie, and J. Zussman 1986. *Rock-forming minerals,* Vol. 1B, *Disilicates and ring silicates.* Harlow: Longman Group UK Ltd.

Dunn, P. J. 1979. "Contributions to the mineralogy of Franklin and Sterling Hill, New Jersey," *Mineralogical Record* 10 (3): 160-5.

Dunn, P. J., P. B. Leavens, and C. Barnes 1980. "Magnesioaxinite from Luning, Nevada, and some nomenclature designations for the axinite group." *Mineralogical Record* 11 (1): 13-15.

Embrey, P. G. and R. F. Symes 1987. *The minerals of Cornwall and Devon.* London: British Museum (Natural History).

Erokhin, Yu. V. and E. S. Shagalov 2007. "Magnesioaxinite from Bakal deposits (South Urals)," *Russian Mineralogical Soc. Proc.* 2007-2-10-0 (in Russian).

Filip, J., U. Kolitsch, M. Novák, and O. Schneeweiss 2006. "The crystal structure of near-end-member ferroaxinite from an iron-contaminated pegmatite as Malešov, Czech Republic," *Canadian Mineralogist* 44 (5): 1159-70.

Foord, E. E., L. B. Spaulding, Jr., R. A. Mason, and R. F. Martin 1989. "Mineralogy and paragenesis of the Little Three mine pegmatites, Ramona District, San Diego County, California," *Mineralogical Record* 20 (2): 101-27.

French, B. M. and J. J. Fahey 1972. "Manganaxinite from the Mesabi Range, Minnesota," *American Mineralogist* 57: 989-92.

Goldschmidt, V. 1913. *Atlas der Krystallformen* Vol. 1 [see Facsimile Reprint in Nine Volumes (1986) by the Rochester Mineralogical Symposium].

Grant, R. and W. E. Wilson 2001. "Famous mineral localities: Dal'negorsk, Primorskiy Kray, Russia," *Mineralogical Record* 32 (1): 3-30.

Hammer, V. M. F and A. Weerth 2004. "Pakistan's Alpine-type clefts: Rivaling the classic Alps," *extraLapis English Edition* No. 6: 72-77.

Hietanen, A., and R. C. Erd 1978. "Ferro-axinites from the Feather River area, Northern California, and from the McGrath and Russian Mission quadrangles, Alaska," *J. Res. U.S. Geol. Survey* 6: 603-10.

Hyršl, J. and Z. Rosales 2003. "Peruvian minerals: an update," *Mineralogical Record* 34 (3): 241-54.

Ito, T.-I., Y. Takeuchi, T. Ozawa, T. Araki, T. Zoltai, and J. J. Finney 1969. "The crystal structure of axinite revised," *Proc. Japan Acad.* 45: 490-4.

Jakob, J. 1923. "Vier Mangansilikate as dem Val d'Err (Kt. Graubunden)," *Schweiz. Mineral. Petrog. Mitt.,* 3: 227-37.

Jobbins, E. A., A. E. Tresham, and B. R. Young 1975. "Magnesioaxinite, a new mineral found as a blue gemstone from Tanzania," *Journal of Gemmology* 14: 368-75.

Kalachev, V. N. 1993. "Axinite: new finds in Russia," *World of Stones* 1: 3-4.

OTHER SCHIFFER TITLES

www.schifferbooks.com

Introduction to Radioactive Minerals. Robert J. Lauf. Collectors have long admired uranium and thorium minerals for their brilliant colors, intense ultra-violet fluorescence, and rich variety of habits and associates. Radioactive minerals are also critically important as our source of nuclear energy. Understanding them is crucial to the safe disposal of radioactive waste. This book provides a 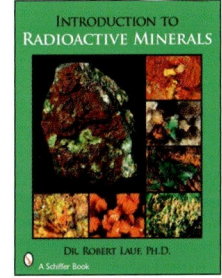 systematic overview of the mineralogy of uranium and thorium, generously illustrated with nearly 200 color photos and electron micrographs of representative specimens. Includes an historical discussion of the discovery of radioactive elements and the development of uranium and thorium ore deposits, a discussion of the geochemical conditions that produce significant deposits, and a description of important localities, their geological setting and history. Major occurrences of interest to mineral collectors are arranged geographically. The minerals are arranged systematically, to emphasize how they fit into chemical groups, and for each group a few minerals are selected to illustrate their formation and general characteristics. With the resurgence of interest in nuclear power, this book is an invaluable guide for mineral collectors as well as nuclear scientists and engineers interested in radioactive deposits.

Size: 8 1/2" x 11"   196 color & b/w photos   144pp.
ISBN: 978-0-7643-2912-8   soft cover $29.95

Collecting Fluorescent Minerals. Stuart Schneider. Seeing fluorescent minerals up close for the first time is an exciting experience. The colors are so pure and the glow is so seemingly unnatural, that it is hard to believe they are natural rocks. Hundreds of glowing minerals are shown, including Aragonite, Celestine, Feldspar, Microcline, 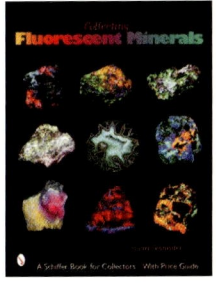 Picropharmacolite, Quartz, Spinel, Smithsonite, plus many more. But don't let the hard-to-pronounce names keep you away. Over 800 beautiful color photographs illustrate how fluorescent minerals look under the UV light and in daylight, making this an invaluable field guide. Included are values, a comprehensive resources section, plus helpful advice on caring for, collecting, and displaying minerals. The field of collecting fluorescent minerals is relatively new and this is one of the most complete references available.

Size: 8 1/2" x 11"   846 color photos   192pp.
ISBN: 0-7643-2091-2   soft cover $29.95

The World of Fluorescent Minerals. Stuart Schneider. The rich and diverse world of fluorescent minerals is explored in this sweeping survey. Breathtakingly pure colors, with their ethereal glow, immediately capture your attention. Did you know that color television is a result of the study of fluorescing minerals? Fresh finds of fluorescent minerals are showing up regularly around the globe, 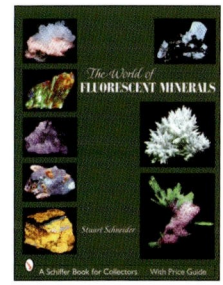 and their collection is an entertaining and popular pasttime. To help the collector, over 825 photos display the minerals both as they might be found in daylight and in under the effects of ultraviolet light. Written for the collector and the merely curious, this pictorial reference will enrich your collecting experience with its informative text. It is an essential source for enjoying and identifying fluorescent minerals.

Size: 8 1/2" x 11"   825 color photos   176pp.
ISBN: 0-7643-2544-2   soft cover $29.95

Schiffer books may be ordered from your local bookstore, or they may be ordered directly from the publisher by writing to:

Schiffer Publishing, Ltd.
4880 Lower Valley Rd.
Atglen, PA 19310
(610) 593-1777; Fax (610) 593-2002
E-mail: Info@schifferbooks.com

Please visit our web site catalog at **www.schifferbooks.com** or write for a free catalog. Please include $5.00 for shipping and handling for the first two books and $2.00 for each additional book. Full-price orders over $150 are shipped free in the U.S.

Printed in China

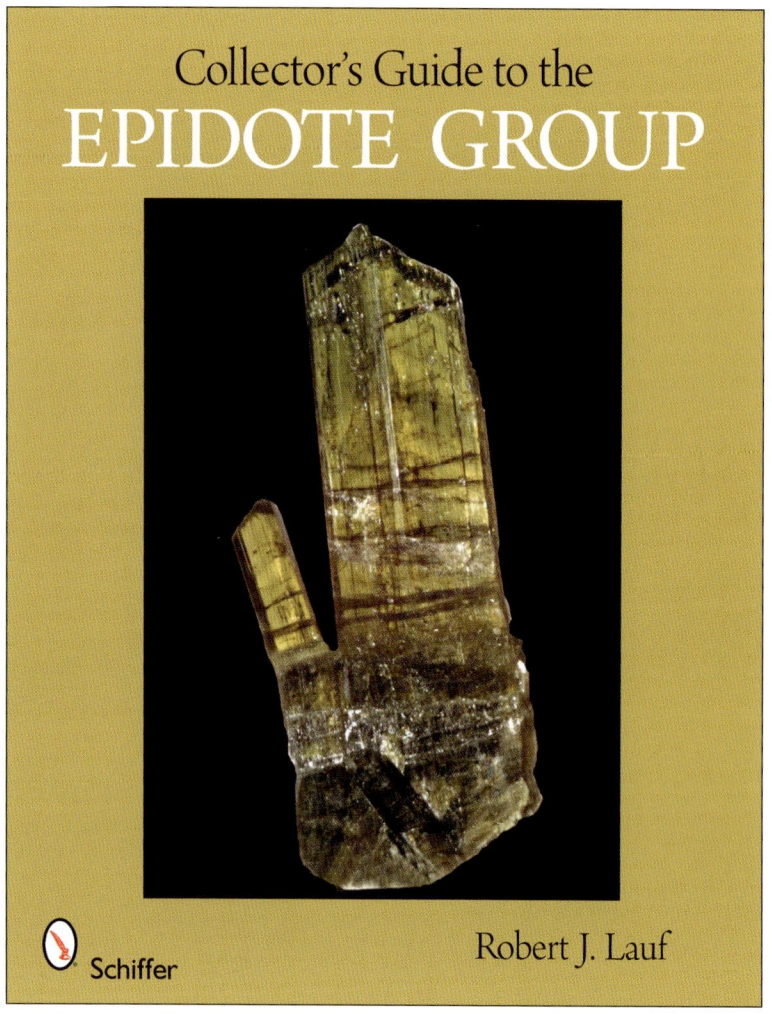

Collector's Guide to the Epidote Group. Robert J. Lauf. Over 90 striking color photos display minerals of the epidote group, well known to mineral collectors for their rich colors and the many interesting minerals with which they occur. Lapidary artists also value epidote, particularly in the form of unakite, and precious or semiprecious varieties of the related mineral zoisite, including thulite and tanzanite, some which have inclusions that allow them to be cut into popular catseyes. This informative book provides all presently known species, detailed entries for each of the eighteen minerals, and extensive locality information. This book will be of interest to those interested in developing a better understanding of silicate minerals.

Size: 8 1/2" x 11" 92 color photos 96pp.
ISBN: 978-0-7643-3048-3 soft cover $19.99

King, V. T. 2003. *Minerals of the USA*, CD version 1.1, Rochester: Rabbit Warren Publishing.

Kunz, G. F. 1892. *Gems and Precious Stones of North America*, Second Edition, (1968 reprint) New York: Dover Publications, 367 pp.

Lauf, R. J. 2007. "Collector's guide to the axinite group," *Rocks & Minerals* 82 (3): 216-20.

Lisitsyn, A. E. and S. V. Malinko 1994. "The Dal'negorsk boron deposit: a unique mineralogical object," *World of Stones* 4: 30-40.

Marchesini, M., and R. Pagano 2001. "The Val Graveglia manganese district, Liguria, Italy." *Mineralogical Record* 32 (5): 349-79.

Milton, C., F. A. Hildebrand, and A. M. Sherwood 1953. "The identity of tinzenite with manganoan axinite." *American Mineralogist* 38:1148-58.

Moroshkin, V. V., and N. I. Frishman 2001. "Dal'negorsk: Notes on mineralogy," *Mineralogical Almanac*, Vol. 4. Moscow: Ocean Pictures, Ltd.

Nekrasov, I. Ya. 1971. "Features of tin mineralization in carbonate deposits, as in Eastern Siberia," *Internat'l Geol. Rev.* v. 13, no. 10, pp. 1532-42. [Trans. from *Sovietskaya Geologiya* 1970, no. 12, pp. 41-54.]

Novak, M., and J. Filip 2002. "Ferroan magnesioaxinite from hydrothermal veins at Lazany, Brno Batholith, Czech Republic." *Neues Jahrbuch fur Mineralogie – Monatshefte*. 9: 385-99.

Peacock, M. A. 1937. "On the crystallography of axinite and the normal setting of triclinic crystals," *American Mineralogist* 22: 588-620.

Pohl, D., R. Guillimette, J. Shigley, and G. Dunning 1982. "Ferroaxinite from New Melones Lake, Calaveras County, California, a remarkable new locality." *Mineralogical Record* 13 (5): 293-302.

Pringle, I. J. and Y. Kawachi 1980. "Axinite mineral group in low-grade regionally metamorphosed rocks in southern New Zealand," *American Mineralogist* 65: 1119-29.

Škoda, R., and R. Čopjakova 2006. "Herzenbergite and Sn-bearing tinzenite from the NYF pegmatite in Třebič pluton, Moldanubicum, Czech Republic," *Acta Mineralogica-Petrographica*, Abstract Series 5: 107.

Sanero, E., and G. Gottardi 1968. "Nomenclature and crystal chemistry of axinites." *American Mineralogist* 53: 1407-11.

Stern, L. A., G. E. Brown, Jr., D. K. Bird, R. H. Jahns, E. E. Foord, J. E. Shigley, and L. B. Spaulding, Jr. 1986. "Mineralogy and geochemical evolution of the Little Three pegmatite-aplite layered intrusive, Ramona, California," *American Mineralogist* 71: 406-27.

Swinnea, J. S., H. Steinflink, L. E. Rendon-DiazMiron, and S. Enciso de la Vega 1981. "The crystal structure of a Mexican axinite," *American Mineralogist* 66: 428-31.

More Schiffer Earth Science Monograph Titles

www.schifferbooks.com

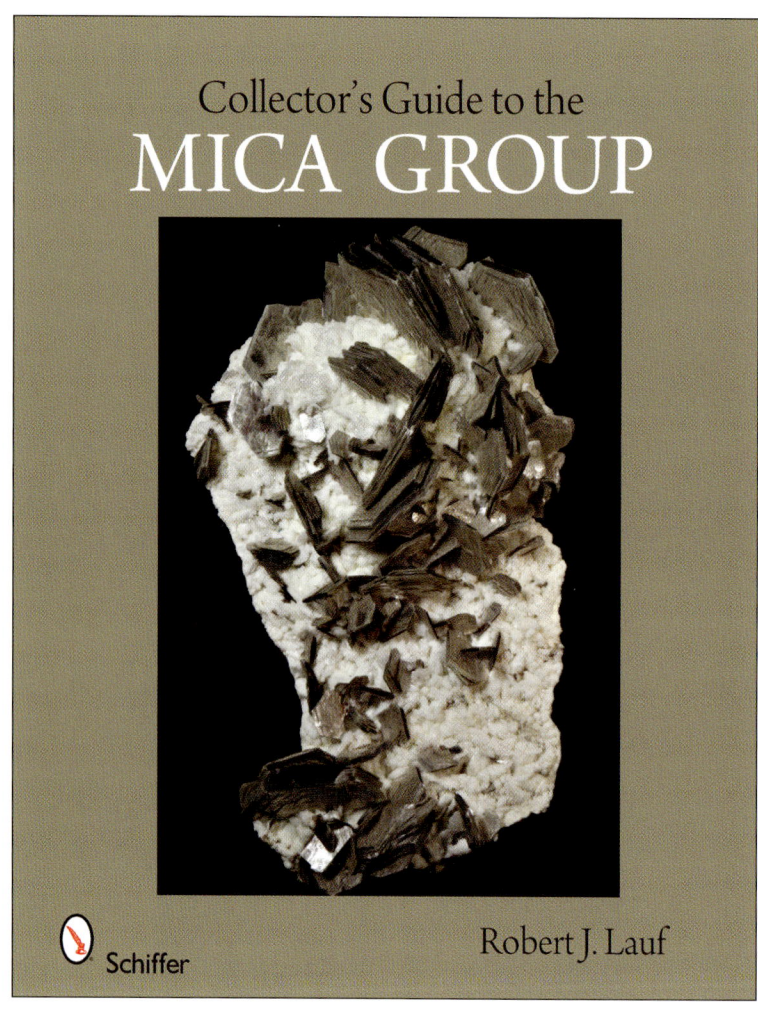

Collector's Guide to the Mica Group. Robert J. Lauf. Mica is a broad term encompassing about forty minerals, ranging from the common to the rare, many at times forming excellent crystals jewelers use. This book feaures examples recently described among the 115 striking color photos and electron micrographs that illustrate the text. A detailed entry for each type includes information on where each is found, associations of micas with other minerals, pseudomorphs (minerals that masquerade as mica), and micas that fluoresce under UV light. This fascinating guide is for those interested in minerals.

Size: 8 1/2" x 11" 115 color photos 96pp.
ISBN: 978-0-7643-3047-6 soft cover $19.99